Hume's Natural Philosophy and Philosophy of Physical Science

Also available from Bloomsbury

Advances in Religion, Cognitive Science, and Experimental Philosophy, edited by
Helen De Cruz and Ryan Nichols
Finding Locke's God, Nathan Guy
Kant's Transition Project and Late Philosophy, Oliver Thorndike
Spinoza in Twenty-First-Century American and French Philosophy, edited by
Jack Stetter and Charles Ramond
The Bloomsbury Companion to Hume, edited by Alan Bailey and Dan O'Brien

Hume's Natural Philosophy and Philosophy of Physical Science

Matias Slavov

BLOOMSBURY ACADEMIC

LONDON · NEW YORK · OXFORD · NEW DELHI · SYDNEY

BLOOMSBURY ACADEMIC
Bloomsbury Publishing Plc
50 Bedford Square, London, WC1B 3DP, UK
1385 Broadway, New York, NY 10018, USA

BLOOMSBURY, BLOOMSBURY ACADEMIC and the Diana logo are trademarks
of Bloomsbury Publishing Plc

First published in Great Britain 2020

Cover design by Charlotte Daniels
Cover image © tmeks/Getty Images

A catalogue record for this book is available from the British Library.

Library of Congress Cataloging-in-Publication Data
Names: Slavov, Matias, author.
Title: Hume's natural philosophy and philosophy of physical science / Matias Slavov.
Description: London ; New York : Bloomsbury Academic, 2020. |
Includes bibliographical references and index.
Identifiers: LCCN 2020026797 (print) | LCCN 2020026798 (ebook) |
ISBN 9781350087866 (hb) | ISBN 9781350185036 (paperback) |
ISBN 9781350087880 (ebook) | ISBN 9781350087873 (ePDF)
Subjects: LCSH: Hume, David, 1711–1776. | Physics–Philosophy. |
Physical sciences–Philosophy. | Philosophy of nature.
Classification: LCC B1498 .S48 2020 (print) | LCC B1498 (ebook) | DDC 192—dc23
LC record available at https://lccn.loc.gov/2020026797
LC ebook record available at https://lccn.loc.gov/2020026798

ISBN: HB: 978-1-3500-8786-6
        ePDF: 978-1-3500-8787-3
        eBook: 978-1-3500-8788-0

Typeset by RefineCatch Limited, Bungay, Suffolk

To find out more about our authors and books visit www.bloomsbury.com
and sign up for our newsletters.

To Emma and Patrik.
Your smiles make my day.

Contents

Figures

Preface

This book centers around David Hume's (1711–1776) natural philosophy and philosophy of physical science. It examines both Hume scholarship and the history of philosophy of physics ranging from Cartesian cosmology to special relativity.

Hume's Natural Philosophy and Philosophy of Physical Science has two equally important aims: First, to deepen our understanding of Hume's relationship to natural philosophy. This aspect has not received considerable attention in Hume studies. Hume is widely acknowledged to be the greatest philosopher to have ever written in English, and consequently today many scholars in the Americas, Europe, Australasia and Japan study his work and publish several academic articles and monographs on it yearly.[1] Second, the book hopes to demonstrate that philosophy and physics have had (and arguably still have) overlapping domains of investigation, including experimentalism, causation, laws of nature, metaphysics of forces, mathematics' relation to nature, and the concepts of space and time.

Acknowledgements

I could not have made it this far without my dedicated PhD supervisors, Mikko Yrjönsuuri and Jani Hakkarainen. Miren Boehm, the opponent in my public defense, was the one to press on me on whether Hume has a natural philosophy to begin with. I hope that, after reading this book, the reader will be persuaded that he does.

Most of this book was written when I was a visiting scholar at UCLA's Department of Philosophy. I am grateful to everybody in the Department. Thanks for the kind invitation by Calvin Normore, approved by Seana Shiffrin, and the funding of The Finnish Cultural Foundation and Alfred Kordelin Foundation, which were both co-ordinated by The Foundations' Post Doc Pool. I wish to thank all the people in the history of philosophy seminar for commenting on my papers: John Carriero, John Kardosh, Graziana Ciola, Antti Hiltunen, Milo Crimi, Aaron West, Janelle DeWitt, Paul Tulipana, and Michael Hansen. I very much enjoyed taking Sheldon Smith's graduate seminar on the laws of nature, and during my stay I benefitted from discussions with him about the philosophy of physics and the philosophy of time. I also found the philosophy of science seminar, organized by Sherrilyn Roush, David Teplow, and Eric Scerri, fascinating and productive. In addition to UCLA, I have had the opportunity to present drafts of my postdoctoral work at: The Socratic Society in Portland State University, The History of Philosophy Roundtable in UC San Diego, Loyola History of Philosophy Roundtable in Chicago, the Scientia conference in University of Minnesota, and the Houston Circle for the Study of Early Modern Philosophy in University of Houston. Many thanks to all the organizers and participants. Concerning the early drafts of this book, I appreciate the advice and critical comments that I received from Don Baxter, Angela Coventry, Dan Kervick, Andrew Janiak, Saul Traiger, Juhana Toivanen, and an anonymous reviewer. All mistakes in this work are mine. As always, my biggest thanks go to my wife. It is difficult to overestimate her contribution. She spent two years in Los Angeles taking care of our son while I did my postdoc. She also read the manuscript of this book and provided valuable feedback. It is to Emma and Patrik that I dedicate this book.

Copyright Acknowledgements

Previous Publications Used in this Book

This book is based largely on my doctoral dissertation, *Essays Concerning Hume's Natural Philosophy* (Jyväskylä: Jyväskylä University Printing House, 2016). I have reproduced material from the following articles:

- "Hume on the Laws of Dynamics: The Tacit Assumption of Mechanism," *Hume Studies*, 42 (1–2): 113–36. © Hume Society.
- "Hume's Fork and Mixed Mathematics," *Archiv für Geschichte der Philosophie*, 99 (1): 102–19. © De Gruyter.
- "Newtonian and non-Newtonian Elements in Hume," *Journal of Scottish Philosophy*, 14 (3): 275–96. © Edinburgh University Press.
- "Empiricism and Relationism Intertwined: Hume and Einstein's Special Theory of Relativity," *Theoria: An International Journal for Theory, History and Foundations of Science*, 31 (1): 247–63. University of the Basque Country (CC BY-NC-ND 2.5).
- "Newton's Law of Universal Gravitation and Hume's Conception of Causality," *Philosophia Naturalis: Journal for the Philosophy of Nature*, 50 (2): 277–305. © Vittorio Klostermann.
- "Ajan havaitsemisesta: Onko aika empiirinen käsite?," In H. Laiho, & M. Tuominen (eds), *Havainto: Suomen Filosofisen Yhdistyksen yhden sanan kollokvion esitelmiä*, 233–40. Reports from the Department of Philosophy, 40, Turku: University of Turku. (CC BY 4.0).

I am grateful to the publishers for permission to use the articles in this book.

Abbreviations

I have abbreviated classical works that I cite throughout the book. The standard abbreviations are indicated in brackets. I have used davidhume.org extensively for studying Hume.

DM   Berkeley, G. (1992), "De Motu," In *De Motu and The Analyst*. ed., trans. D. M. Jesseph, 73–107. Dordrecht: Kluwer Academic Publishers.

DNR   Hume, D. (2007), *Dialogues Concerning Natural Religion*, ed. D. Coleman, Cambridge: Cambridge University Press.

EHU   Hume, D. (2006), *An Enquiry Concerning Human Understanding*, ed. T. L. Beauchamp, New York: Oxford University Press.

EMP   Hume, D. (1998), *An Enquiry Concerning the Principles of Morals*, ed. T. L. Beauchamp. Oxford: Oxford University Press.

Essay   Locke, J. (1996), *An Essay Concerning Human Understanding*, K. Winkler (ed.), Indianapolis: Hackett Publishing.

History   Hume, D. (1983), *The History of England*, ed. W. B. Todd, 6 vols, Indianapolis: Liberty Classics.

KdRV   Kant, I. (1998), *Critique of Pure Reason*, P. Guyer, A. Wood (trans), New York: Cambridge University Press.

L   Hume, D. (1967), "A Letter from a Gentleman to his Friend in Edinburgh," ed. E. C. Mossner and J. V. Price, Edinburgh: Edinburgh University Press.

Opticks   Newton, I. (1952) *Opticks or A Treatise of the Reflections, Refractions, Inflections and Colours of Light*, New York: Dover Publications.

Principia   Newton, I. (1999), *The Principia: Mathematical Principles of Natural Philosophy*, trans. I. B. Cohen, A. Whitman, and J. Budentz, Los Angeles: California University Press.

Principles   Descartes, R. (1983), *Principles of Philosophy*, trans. V. R. Miller and R. P. Miller, Dordrecht: Reidel.

T   Hume, D. (2000), *A Treatise of Human Nature*, ed. D. F. Norton and M. J. Norton, New York: Oxford University Press.

Introduction

This book will introduce a new David Hume: Hume the natural philosopher. I am by no means claiming that Hume is first and foremost a natural philosopher. He is not; for his main objective is to establish a new science of human nature. Hume did not provide metaphysical foundations for physics, as did his contemporaries Du Châtelet and Kant, for example. However, his work in many ways engages in the tradition of natural philosophy. This aspect of Hume has not been given the attention it deserves.

This work is not the first to examine Hume on the topic of natural philosophy, or the history of physical science. In 1930 Mary Shaw Kuypers published *Studies in the Eighteenth Century Background of Hume's Empiricism*. In the Preface, Kuypers ([1930]1966) notes that Hume's philosophy has been treated "chiefly in its relation to the English epistemological tradition." She continues: "There has been, however, no detailed study of the relevance which Hume's philosophy may have had to the problems which science has raised." Although Kuypers considers the same issues as I do in this book, a fresh look at them is justified. For example, section two of Kuypers' book focuses on the interpretations of Newtonian science in the eighteenth century. Since 1930, there have been many studies of Scottish enlightenment scientific culture (of which Barfoot's 1990 study is a paradigmatic example). Newton scholarship has increased rapidly in recent years, and we now have a different picture of Newton after John Maynard Keynes purchased his private manuscripts from a Cambridge auction in the mid-1930s, and "De Gravitatione" was published in 1962. Moreover, plenty of Hume scholarship has been produced since the 1930s. Our picture of Hume's philosophy of causation is now richer, thanks to interpretations other than the traditional regularity reading, such as projectivism (e.g., Beebee 2006), quasi-realism (e.g., Coventry 2006), and the causal realist reading (e.g., Kail 2007).

There are two recent books that treat Hume's background in early modern natural philosophy that I am aware of: *The Natural Origins of Economics* by Margaret Schabas (2005), and *David Hume and the Culture of Scottish*

Newtonianism by Tamás Demeter (2017). The two have slightly different angles compared with this book. Schabas focuses on the natural philosophical foundations of Hume's political economy and Demeter concentrates on the background of Hume's moral philosophy in the *Opticks*-inspired tradition of Scottish vitalist natural philosophy. I hope I can fill the gap for a book on Hume's relation to natural philosophy and philosophy of physical science.[1]

In thinking of a title for this book (and the dissertation it is based on), I wanted to stress the historical concept of "natural philosophy." I consider the umbrella term "natural philosophy," or "philosophy of nature," to be a more accurate depiction than the contemporary notions of "epistemology" or "philosophy of science." I do not claim that the latter contemporary notions are illegitimate. Specific parts of this book are also about epistemology and philosophy of science, as well as about metaphysics and philosophy of mathematics, and so on. However, there are important reasons to think that the historical term "natural philosophy," which Hume himself widely applies,[2] encapsulates the overall topic of the book better than "epistemology" or "philosophy of science." The problem with the label "epistemology," understood in the pre-Quinean sense, is that it presumes a distinction between a philosophical, often normative, theory of knowledge and an empirical psychological investigation of the human mind. However, nowhere in Hume does this kind of distinction appear explicitly. Hume did not use the terms "epistemology" or "psychology" himself. The term "epistemology" was used for the first time by James E. Ferrier in 1854 (Woleński 2004: 3). In the eighteenth century, what we now call psychology was placed under the label of "moral philosophy." Thus one could say that, to use our contemporary constructs, Hume's philosophical project employs both psychology and epistemology. Hume's theories of perception, memory, imagination, and personal identity in the first book of the *Treatise* could very well be conceived as works of cognitive psychology (or its proto-form). In turn, his distinction between "proofs" and "probabilities" in the tenth section of the first *Enquiry* utilizes normative epistemic standards, as Hume recommends that "a wise man" should proportion "his belief to the evidence." In this sense, Hume is not only interested in describing the way human cognition works but also in prescribing epistemic virtues which should be adopted by a reasonable cognizer.

The problem with the contemporary notion of "philosophy of science," when applied to Hume's natural philosophy and its context, is that by the very definition of the term, it presupposes a categorical distinction between "science" and "philosophy": "philosophy of science" is philosophizing about science. Such a categorical distinction cannot be accurately applied to the early modern intellectual world. This would give us too narrow a picture of what natural philosophy is about. To explain this (here only briefly; a more comprehensive treatment will be provided in Chapter 1), it is useful to refer to Andrew Janiak's (2015: 18–19) study on the matter: "Seventeenth-century philosophers who studied nature investigated such things as planetary motions, the nature of matter, causal relations, and the possibility of a vacuum, but they also discussed many aspects of human beings, including the human psyche or the soul, and also how nature reflects its divine creator."

We may consider, for example, Newton's *Principia: Mathematical Principles of Natural Philosophy*, which arguably is the major natural philosophical work of the era. It includes aspects that in contemporary language we can understand to be physics, such as an axiomatic system based on mathematical definitions and propositions concerning laws of nature, and the application of computational, observational, and experimental techniques. The mathematical-empirical inquiry of the work can be properly labeled as "science," to use our contemporary language. But some parts of the *Principia* can be understood as being closer to what we would call philosophy. Newton's dynamics lean heavily on a philosophical notion of causation, and his argument for absolute space and time, which is intended to give his laws of motion a robust realist status, is clearly a metaphysical-philosophical pursuit. The Introduction, the section of Rules for the Study of Natural Philosophy, and the General Scholium in the second edition of the work include methodological remarks and critical thinking about the foundational epistemological and ontological issues concerning experimentation, induction, explanation, the universality and the reliability of results, mechanism, and intelligibility. These aspects of the *Principia* could be considered as "philosophy." As the paradigmatic work of natural philosophy does not contain any dichotomous distinction of science and philosophy (or theology, for that matter),[3] we may say that in the early modern world there was no sharp dividing line between the two disciplines, as we might understand the difference today.[4]

Method and Viewpoint

Quine has been reported as saying that "there are two sorts of people interested in philosophy, those interested in philosophy and those interested in the history of philosophy" (MacIntyre 1984: 39–40). This witty remark was probably intended as a joke. But is such a distinction feasible?

For the sake of this argument, Robert Piercey (2003) separates the ideal philosopher who is doing pure philosophy and the historian of philosophy, who is merely interested in the thoughts of authorial figures of the past. Unsurprisingly, he (ibid.: 783) holds this dichotomy untenable:

> The pure philosopher, as I have described him, is interested solely in the answers to philosophical questions, not in their history. But it is obviously impossible to try to answer philosophical questions until one has learned "what questions are the genuinely philosophical ones". And this, surely, is something one learns largely through an acquaintance with history—by seeing which questions philosophers have traditionally posed, how these questions differ from those traditionally posed by other enterprises, and so on. Likewise, the pure historian of philosophy, as I have described him, wants to understand past philosophers in their own terms, rather than filtering their work anachronistically through contemporary concerns. But does this goal even make sense?

Quine's purported joke reveals an impossible dichotomy. One cannot be a "pure" philosopher, because what counts as philosophy is shaped by an earlier tradition. Current issues in systematic philosophy have a history of thousands of years. A good example comes from the philosophy of time. Broadly speaking, McTaggart's (1908) paper defines the options for contemporary metaphysics: either A- or B-theory is thought to provide the correct metaphysics of time. To put this starting point succinctly, we cannot accept both past–present–future dynamic presentism and before–after static eternalism (Slavov 2019). So, contemporary philosophers of time pursue either A- or B-theory. But McTaggart did not invent the original distinction. Among the Presocratics, Heraclitus stated that everything is in constant flux, and Parmenides argued that change is unreal. This is somewhat analogous to the contemporary debate between the dynamic and static positions, although contemporary B-theorist do not dispute the reality of change (Deng 2018). Importantly, the very reason that the plethora of issues in systematic philosophy of time count as philosophical is because these are perennial issues in philosophy. One cannot do "mere" history of philosophy, either. The historian of philosophy has some philosophical interest in writing a piece of history of philosophy. Even simple tasks like quoting or choosing a section from a book to be perused are based on a specific interest. Without philosophical engagement in historical scholarship, the result would be mere duplication of older texts (Rorty et al. 1984: 11).

Philosophical, not merely exegetical, involvement with the texts is also apparent in the practice of doing history of philosophy. It is desirable to produce a novel interpretation, to show that it is consistent, and to defend it against other interpretations. This is clearly philosophical work. However, history of philosophy differs from systematic philosophy in that it typically assumes the authorial nature of historical figures. The task is to understand and to expound on the views of a classical thinker, not to debunk or vindicate them. In this book, I assume this stance for the most part, too. Chapter 7, the final chapter, is an exception. In it I will not only examine Hume's views, but will assess Humean philosophy critically when it comes to the relativity of time, metaphysics of laws of nature, and the relation of causation and physical laws.

The method used in this book is problem-oriented contextualization of central arguments. By textual I mean original publications of historical authors, their correspondence, and the manuscripts they may not have published themselves, as well as contemporary secondary sources. Although the chapters focus on Hume's concept of natural philosophy, his views are compared with other contributions to the history of philosophy and science. When referring to the historical context of a philosophical or a scientific argument, I have in mind the textual environment, rather than the non-textual sociocultural environment. Tad M. Schmaltz (2013: 319–20) clarifies this distinction:

> History of philosophy, as a branch of philosophy, focuses primarily on the philosophical context of a particular text, as provided in other texts. In contrast, social history of science, as a branch of history, is concerned primarily with the non-textual environment of most interest to historians. Of course, this is to say neither that philosophical historians of philosophy are uninterested in the social context, nor that social historians of science are uninterested in the philosophical context. It is a matter rather of where the priorities lie: is the philosophical upshot of the text the main point, or is the concern rather with social connections and implications?

My priority is to analyze and interpret "the philosophical upshot of the text" and its context understood in the above-mentioned sense. The context of an argument refers to the surrounding philosophical and scientific works (see also Galison 2008: 113). My approach to examining the texts borrows from argument analysis. The basis of justification, background assumptions, and consistency are the focus. Hume's arguments are placed into a broader philosophical and scientific framework, and interpreted in a historically sensitive way. I hope this approach offers both a systematically rich and historically informed interpretation of Hume's work.

Here is a specific example of this book's approach. In Chapter 7 I explore Hume's impact on subsequent science. The example I have chosen is Einstein's special theory of relativity. Einstein read Hume shortly before devising the theory, and implemented Humean concept empiricism in his argument for the relativity of simultaneity. The denial of absolute simultaneity, or absolute time, enabled him to reconcile the two seemingly irreconcilable aspects of the theory, that is, the principle of relativity and the light postulate.

Such an approach is likely to face objections. One could say that there are many, many more issues involved in the formulation of the theory: industrialized society which utilized electromagnetic induction in generating energy; pressure to improve clock synchronization due to increasing train traffic; a long history of physics ranging from Young's diffraction experiments to Maxwell's equations with all its intricate mathematics. A huge variety of social, technological, scientific, and mathematical factors played a role in Einstein's discovery. In my analysis, I am deliberately concentrating on philosophical matters. Even the validity of this viewpoint can be challenged. The philosophical views preceding and associated with relativity are not solely Hume's. Leibniz and Berkeley criticized Newtonian absolute conceptions of space and time before Hume's contributions. Similar views were also presented by Einstein's contemporaries, such as Mach and Poincaré. I believe one could find several other philosophers/philosophies that are congenial to the ramifications of relativity.

I acknowledge that all such vantage points are relevant—and doubtless, there are many other cogent perspectives one could take concerning the history and philosophy of special relativity. My response to the methodological challenge is to simply note that I have chosen a specific standpoint. I look at the connection between Hume's philosophy and special relativity from an epistemic and ontic perspective. Relevant issues for comparison between a philosophical doctrine and a scientific theory involve questions like: What is the origin of our concepts of space and time? What are their points of reference, and how are they justified? In which sense do space and time exist?

Chapter Outlines

Chapter 1 tackles the definition of natural philosophy, both systematically and historically. Central to this task is to delineate the difference and overlap of physics and philosophy. It is argued that there is a gray area between the two, and that this is the proper field for natural philosophy. The argument is bolstered by a systematic consideration of a paradigm case of natural philosophy, the notion of a law of nature, and then by a brief historical overview of the central results of Newton's *Principia*.

Chapter 2 analyzes the relation between Hume's science of human nature and natural philosophy. It is agreed that Hume is first and foremost a scientist of the human mind. Still, a considerable amount of his work engages with and is significant to natural philosophy. This will be shown by both problem-oriented contextualization and referring to Hume's actual education in natural philosophy. Moreover, Hume's partial criticism of metaphysics will be compared with his partial criticism of natural philosophy, and it will be concluded that Hume has some natural philosophy in a similar way as he has some metaphysics.

Chapter 3 traces Hume's experimental commitments back to the views of his fellow Britons, Boyle and Newton. Regarding Boyle, the focus will be on his social-epistemological arguments that require recurring witness testimonies to experiments. In Newton's case, issues related to his methodological defense of the law of universal gravitation against the charges of speculative philosophy will be highlighted. Hume's notion of experience is compared to the notion of experiment, and the nature of Humean induction as extrapolation from the observed to the unobserved is examined.

Chapter 4 begins by clarifying the notion of a law of nature in its early modern context. It will be argued that Hume subscribes to a bottom-up analysis of laws. In accordance with the traditional regularity reading, it is argued that for him causation is a constant conjunction of species of objects and events which we identify with experience. An original reading is set forth as it is emphasized that for Hume constant conjunctions are discovered in nature.

Chapter 5 delves into Hume's distinction between relations of ideas and matters of fact. The dichotomy of mathematical and empirical propositions is elucidated, and it is argued that propositions of mixed mathematics are empirical for Hume. Accordingly, mathematics does not concern nature directly. To make this point more convincing, it is added that Hume remains agnostic on whether nature has an underlying mathematical (for example, geometric) structure. Our idea of extension is acquired from the senses of vision and touch, and without the corresponding sensory qualities we cannot comprehend, or even think of, reality itself as mathematical.

Chapter 6 concentrates on the concepts of space and time. Newton's absolutism is contrasted with Descartes' relational approach. It is pointed out that Hume is closer to Descartes than Newton in his views on space and the vacuum, and that Hume eschews absolute universal Newtonian time.

Chapter 7 starts by inspecting the relation of Hume's philosophy to Einstein's special relativity. The views of the philosopher and the physicist are juxtaposed. It is argued that Einstein's argument for the relativity of simultaneity is hospitable to Humean concept empiricism, and in tension with Newton's absolutist and Kant's transcendental arguments. Nevertheless, Hume's radical, and therefore skeptical, empiricism is not consistent with timelike separation of special relativity, which requires substantial existence of physical events. The rest of the chapter surveys Humeanism in contemporary philosophy of physics, on the topics of laws of nature, and their putative causal nature. The final subsection briefly considers the Humean-inspired notion of unintelligibility in relation to modern physics.

All the chapters begin by introducing the topic under discussion. This suits readers who are not well acquainted with the concepts and theories involved. Consequently, the objective of each topical section is to provide an original interpretation of the relation between Hume and natural philosophy.

The Concept of Natural Philosophy

This chapter focuses on the concept of natural philosophy. I provide both a systematic and a historical account of what natural philosophy is. With regard to the historical account, I limit myself to explicating how natural philosophy was understood in the early modern period. A critical aspect in defining natural philosophy is to grapple with the relation of philosophy and physics.[1] The two are different, but a difference does not imply a dichotomy. I believe there to be a gray area between philosophy and physics. This gray area, in which the two domains of investigation overlap, is the proper field for natural philosophy.[2]

My examination of natural philosophy in this chapter runs chronologically backwards. I wish to start by examining how we might think about the relationship between philosophy and physics today. I have a specific rationale for choosing a reverse chronological order. Philosophy and physics clearly make up different domains of research for us twenty-first-century people. This might lead us onto the wrong track right from the beginning. We could fail to see the common issues that both philosophy and physics touch upon. A systematic definition of the concept of natural philosophy brings down a putative dichotomy (but not a difference) between the two, and therefore enables us to see better that these disciplines have many common intellectual roots. I understand this systematic approach as being necessary for assessing the relevant history of natural philosophy.

Philosophy and Physics

Philosophy and physics are nowadays very different branches of study. I will attempt to explain this difference succinctly. To do this, I have to both simplify and generalize. Simplification and generalization are unavoidable because there are so many different kinds of philosophies, as there are also many sub-disciplines within physics.

Here is a very basic introduction. Philosophy is involved with dialogic argumentation, *a priori* reasoning, intuition, formal logic, and conceptual analysis concerning the most fundamental questions. Philosophers tackle issues such as what exists, and how we come to know of its existence. Physics, for its part, investigates quantifiable natural phenomena with mathematics, by means of observation and experimentation. Physicists solve problems involving, for example, the relations of matter, motion, time, and energy. The methods of study and the specific problems addressed are not the only differences among the two disciplines. There is a major institutional gap between the two as well. Both in the secondary school teaching and in the academia, philosophy finds its home in the humanities and the social sciences, whereas physics stands as the paradigmatic example of natural science. Due to increasing specialization since the nineteenth-century industrial revolution in Western societies, a renowned expert in one field might be a complete amateur in another. In explaining this social divergence, C. P. Snow ([1959]1998) famously stated that the humanists/social scientists and the natural scientists make up "two cultures" of their own.[3]

For this book, it is vital to examine the philosophy/physics dichotomy. This does not mean that philosophy and physics are the same thing. They are not, for the reasons—different methodologies, specific research questions, and social institutions—I sketched above. But differences do not equal a dichotomy. As this is such a crucial starting point for this book, I wish to expound on this point in detail.

Consider the following proposition, which is a variant of Newton's second law of motion: "It is a law of nature that when an object is impressed by a force, it will change its state of motion." There is no doubt that this is a scientific proposition.[4] To test the proposition is to engage in physics. In its most generic form, the sample proposition can be expressed in terms of mathematics, $\vec{F} = \dfrac{d\vec{P}}{dt}$, where $\vec{F}$ is the force acting on the object, $\vec{P}$ is the total linear momentum, which is defined as the multiplication of the mass, m, and the velocity, $\vec{v}$, of the object, and t denotes time. A modified version of the proposition is subject to a simple experiment. Assume that mass is constant, so we can deduce that the general proposition in the form of $\vec{F} = m\dfrac{d\vec{v}}{dt} . \dfrac{d\vec{v}}{dt}$ is a definition of change in motion, that is, acceleration, so the proposition can be expressed as $\vec{a} = \dfrac{\vec{F}}{m}$, in which $\vec{a}$ stands for acceleration. Now we can create a scenario involving a low friction object that is pulled by variable weights, and measure the distances it covers in a given time under the influence of varying forces. Consequently, it can be shown, within a margin of error that the force exerted on the object is directly proportional to the acceleration produced, that is, $\vec{F} \propto \vec{a}$.

At first, we may gather that the sample proposition, "It is a law of nature that when an object is impressed by a force, it will change its state of motion," is all about physics. Examining the proposition requires defining physical quantities, mathematical derivation, creating an idealized target system that neglects factors like friction and air resistance, carrying out the experiment for a number of times, and finally presenting the results to peers in a quantitative and a graphical form. The proposition is an item of rudimentary physics, but, all things considered, it is a part of science.

Given the former exposition of a proposition of physics, one might be left wondering: What on earth could philosophy have to do with any of this? Philosophical ambition of addressing foundational questions about what exists and how we come to know it seems to be as far away from physics as it gets.

Still, a perusal of the sample proposition reveals that one may adopt a philosophical perspective to it, too. The proposition mentions a law of nature. What is a law of nature? Are laws of nature contingent regularities, like constants of nature such as friction and air resistance, or are they something more; do they instantiate physical necessity? This leads us to the question of what kinds of modalities pertain to the world: contingency, physical necessity, and/or logical necessity? The proposition is expressed in causal terms, as it mentions that an object is "impressed" by a force. Are laws, then, causal? If they are, what objects or events described by the proposition should be considered as causes and effects? What are objects or events, precisely? What kind of criteria should our judgments concerning causes and effects satisfy? Should cause and effect be contiguous, temporally successive, and distinctly separable? If these requirements are not satisfied, should we then conclude that laws do not instantiate causation, but rather that laws enable us to predict the probable outcome of an isolated system? How is it possible to identify causation? Is it by means of observing regularities, or by considering counterfactual terms, or by manipulating salient variables? The proposition refers to the term force, but later theories in fundamental physics do not make a reference to Newtonian forces. In what sense, then, do forces exist? Are they merely fictitious entities that are useful for engineering purposes, or do they exist in some effective, non-fundamental way? Do scientific propositions express historically specific and ever-changing ways of making sense of the phenomena around us, or is science able to grasp the structure of the world as it really is in itself? Is there a deep incongruity of different world views between the old and the new scientific theories, or do new ones contain an element of the old ones? And what about the role of mathematics in physics: is it just a calculating device for making satisfactory predictions, or does nature itself have a structure that can be identified with the aid of mathematics? And, finally: should metaphysics constrain physics in deciding these questions, or should it be the other way around?

The former analyses of the interrelated semantic, epistemic, ontic, and metaphysical issues indicate that the sample proposition relates to philosophy. Although the proposition is doubtlessly a physical proposition, it is imbued with philosophy. Thinking about modalities, causation, the structure of reality, and our limits of knowing it is philosophical activity *par excellence*.

Again, my purpose is not to demonstrate that philosophy and physics are the same enterprises. They are not. But there is something like a gray area between philosophy and physics. In my understanding, this gray area is the proper field for *natural philosophy*. Defining this term is not a straightforward issue. It could be defined as an obsolete usage of the word physics. There is an element of truth to this. In the early modern period, natural philosophy was essentially understood as a discipline that is involved with issues that we now call physics (and which the early moderns themselves sometimes referred to as "physicks," too). I shall analyze this historical aspect in the next section of this book. Before that, I wish to make the notion of natural philosophy clear from a systematic perspective. Here I shall rely on Lee Smolin and Roberto Mangabeira Unger's (2015: 75–7) definition. In their account there are four key elements that natural philosophy is composed of. In the following, I provide my interpretation of Smolin and Unger's stipulations.

First, the subject matter of natural philosophy is nature. This is different from philosophy of science, because its subject matter is science itself. Surely natural philosophy is interested in the nature of science, like the structure of scientific theories, scientific inference, and scientific practice. Natural philosophy is not miles away from the philosophy of science (which is why I wanted to include the term philosophy of science in the subtitle of this book). But this interest is based on the larger argument of what nature is like, given the constraints of our limited cognitive capacities.

Second, natural philosophy has a critical objective. It aims to dissociate the scientific discoveries of nature from the interpretation of scientific theories. It does this from a distance; natural philosophy is not a part of science in the sense that it carries out empirical research or devises hypotheses for testing new theories. Rather, natural philosophy explores the semantics, epistemology, ontology, and metaphysics that relate to what physics has to say about nature. This can be beneficial for a new science in its early stages.[5] A good example is the formulation of special relativity in the early twentieth century. The creation of the theory involved a philosophical evaluation of the concepts of space and time.[6] Although special relativity ensued from a critical reflection of nineteenth-century electrodynamic physics and the mathematics related to it, philosophy was of central importance to the conceptual revisions of the notions of space and time.

Third, although natural philosophy maintains some critical distance to science, it does not set forth views that are independent of or above science. Natural philosophy has an interest in figuring out what nature is like by paying close attention to what physics has to say about it. Natural philosophy does not unravel facts about the world that would be somehow deeper than what physics may accomplish. This makes natural philosophy deviate from aprioristic metaphysics (as championed by, for example, MacDonald (2005), Lowe (2011), and Fine (2012)), which holds that metaphysics can establish knowledge about reality that is more fundamental than a scientific description of it. However, natural philosophy comes close to a naturalistic metaphysics (as envisaged by, for example, Maudlin (2007), Ladyman and Ross (2007), and Maclaurin and Dyke (2012)) which emphasizes that there is a continuum between philosophy and the sciences.[7]

Fourth, as Smolin and Unger (2015: 77) put it, natural philosophy "intervenes in discussion of the agenda of natural science," by attenuating "the clarity of the divide between a discourse within science and a discourse about science." So natural philosophy has a critical task in figuring out the limits of scientific inquiry. It neither just blindly follows the results of the natural sciences nor claims to have superior knowledge of nature compared with scientific results.

The systematic definition of natural philosophy that I have made in this section is, I shall argue, also important for understanding early modern natural philosophy. In the aftermath of Newton's *Principia*, philosophy and physics started to separate. Still, scholars of the period were very much intrigued by the gray area between philosophy and physics, that is, by natural philosophy.

Early Modern Natural Philosophy

Ephraim Chambers' encyclopedia from the year 1728 (617) defines "natural philosophy" as: "Natural Philosophy, that Science which considers the Powers of Nature, the Properties of Natural Bodies, and their mutual Action on one another; otherwise call'd Physicks. See Physicks." Chambers equates natural philosophy with the science of physics. His definition of "physicks" further divides, "with regard to the manner wherein it has been handled, and the Persons by whom, into 1° Symbolical, [...] 2° Peripatetical, [...] 3° Experimental, [...] 4° The Mechanical or Corpuscular."

On first reading, as with the sample proposition provided in the previous section, Chambers' definition might lead us to wonder whether natural philosophy had anything to do with philosophy even in early modern period. We might look into Newton's *Principia – The Mathematical Principles of Natural Philosophy*,[8] which is arguably the paradigmatic work of natural philosophy of its time. In the book, we find an investigation of natural phenomena, namely motions of objects under the influence of forces, with mathematics and by observational and experimental techniques. To show that Newton's *Principia* is essentially physics as understood by our contemporary vocabulary, it is important to take a brief sojourn to its central results.

Newton begins his *Principia* with the section "Definitions." He first defines mass, which he calls "quantity of matter," to emphasize the fact that mass is invariable. In this sense, mass is the bulk of matter, unlike its variable weight: "Quantity of matter is a measure that arises from its density and volume jointly" (Definition 1). Then he goes on to define linear momentum, which he calls "quantity of motion," as follows: "Quantity of motion is a measure of motion that arises from the velocity and the quantity of matter jointly" (Definition 2). After this he adds another definition of mass: it is the resisting factor of change in motion, to wit, inertia (Definition 3). The fourth definition explicates the notion of a force: "Impressed force is the action exerted on a body to change its state either of resting or of moving uniformly straight forward" (Definition 4). The rest (Definitions 5–8) concern the centripetal concept of force.

After the Scholium to the Definitions Newton proceeds to an axiomatic exposition of the laws of dynamics. In his treatment of these laws, there is one major difference compared with a contemporary textbook presentation of dynamic laws that I provided in the previous section; Newton does not employ calculus but Euclidian geometry. For him instantaneous impulsive forces are primary, whereas continually acting forces are secondary.[9] His second law of motion states: "A change in motion is proportional to the motive force impressed and takes place along the straight line in which that force is impressed." If a body is impressed by multiple instantaneous forces, the resultant uniform motion takes place in the direction of the sum of the forces. Newton depicts this geometrically. If force *M* is impressed from *A* to *B*, and force *N* from *A* to *C*, the resultant inertial motion happens in the direction of the diagonal of a parallelogram, from *A* to *D*. Newton images the resultant motion in Figure 1:

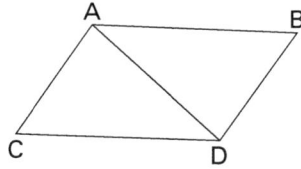

Figure 1 Newton's parallelogram of forces in the *Principia*.

Newton's definitions and his laws figure prominently in his interpretations of astronomical observations. They are necessary for his comprehensive argument for the law of universal gravitation (henceforth LUG). Newton's argument consists of three main parts (Belkind 2012; Harper 2016). The first major step of the argument can be found in Book 1, Section 2 of the *Principia*. In Proposition 1, Newton provides an idealized figure of Kepler's area law. In the figure he depicts a body subjected to instantaneous forces along a polygonal trajectory:

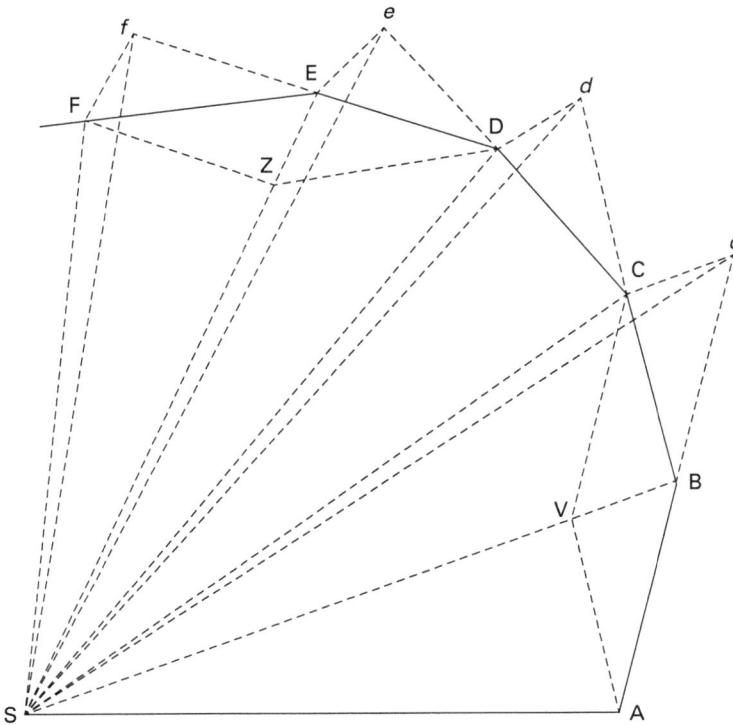

Figure 2 Demonstrating Kepler's area law and the concept of centripetal force.

When moving around an immobile center, the body sweeps equal areas in equal amounts of time. Based on the area law, Newton provides a geometric proof by reasoning counterfactually; if there were no change of motion in the direction of the immobile center, the body should move rectilinearly (from A to c instead of from A to C). Accordingly, Propositions 2 and 3 evince that the centripetal force of an orbiting body is directed toward an inertial center.

In the next step of his argument, Newton employs Kepler's harmonic rule. The rule states that the square of the orbital period of a body around the immobile center is directly proportional to the cube of the semi-major axis of its orbit. In Proposition 4 Newton approximates that satellites move circularly instead of elliptically. With this idealizing condition he deduces from the harmonic rule the following conclusion: the centripetal acceleration of the orbiting body is inversely proportional to the square of distance from the center of orbit.

Before the last step of the argument for LUG in Propositions 6 and 7 in Book 3 of the *Principia*, Newton establishes in the preceding Scholium[10] that gravity is a centripetal force: "Hitherto we have called 'centripetal' that force by which celestial bodies are kept in their orbits. It is now established that this force is gravity, and therefore we shall call it gravity from now on."

In Proposition 6 Newton describes pendulum experiments he carried out with nine different materials. The invariant periods of oscillations with the different materials in the experiment confirm (within a minuscule margin of error) that the gravitational acceleration is independent of the mass of the falling object near the surface of the earth. Earlier in Proposition 4 Newton had shown that the moon's centripetal acceleration toward the earth is dependent on its distance. This calculated acceleration is consistent with the inverse-square magnitude of gravity in the same measure as with objects near the surface of the earth. "Therefore," in his own words, "that force by which the moon is kept in its orbit, in descending from the moon's orbit to the surface of the earth, comes out equal to the force of gravity here on earth." This ties together Newton's two earlier argumentative steps in Propositions 1–4 of the first Book, which applied both the centripetal concept of force and the inverse-square measure of distance of that force. He then continues to universalize the domain of gravitational force in Proposition 6. Jupiter's moons accelerate toward Jupiter with the magnitude that is only dependent on their distance from the planet's center. This follows both from the harmonic rule and, respectively, from the area law. Accordingly, all primary planets accelerate toward the sun.

In Proposition 7 Newton makes the sweeping claim that the force of gravity "is proportional to the quantity of matter in each." He adds in Corollary 1 of the Proposition that as the bulk of the matter attracts, all mass points (instead of the surfaces of bodies) attract: "For the force of the whole will have to arise from the forces of the component parts." The result of Newton's argument is that every body in the universe attracts any other body with a force that is directly proportional to their masses and inversely proportional to the square of their distance.

Given the core results of the major work of early modern natural philosophy, and natural philosophy's dictionary definition, which expressly says that it is the science of physics, it is legitimate to ask: What role does philosophy have in any of this? It seems that natural philosophy, already in the early modern era, was all about physics. The scientific aspects of the work are, broadly speaking, a match with our understanding of what science is. We even find support for this view from a contemporary encyclopedia article on the notion of *scientiae* in renaissance (in which renaissance denotes roughly the same period as early modern). Tamás Demeter, Benedek Láng and Dániel Schmal (2015: 2) agree with the match between science and natural philosophy, as they argue that "what we understand by the term 'science' today, that is, the systematic investigation of nature, would be better described in the Renaissance by the term *philosophia naturalis*, natural philosophy." They add that "this is not coextensive, however, with the present-day scope of science, because to the category of natural philosophy 'mathematical sciences' should also be added." This addition further amplifies the point that Newton was doing what we now call science, because his *Principia* highlighted a *mathematical* treatment of natural phenomena. He himself exalts the reduction of "the phenomena of nature to mathematical laws" (Newton 1999: 381).

As noted by Andrew Janiak (2015: chapter 2), Edward Grant (2007) and I. Bernard Cohen and George E. Smith (2002) advance this line of thinking further. Grant (2007: 314) notes that:

In judging what Newton was really doing in his *Mathematical Principles of Natural Philosophy*, we should not be misled by the title. [. . .] Whatever he might have named his treatise, Newton was mathematizing natural philosophy and, depending on the subject matter, also was producing particular sciences. Whatever terms he may have used in his title, a glance at the approximately 530 pages of the *Principia* can leave no doubt that Newton was doing mathematical physics.

Cohen and Smith (2002: 2) make a similar kind of observation. They claim that it is "superficial" to think that:

> what we now call science was then still part of philosophy, so-called "natural philosophy" as in the full title of the work that turned Newton into a legend, *Philosophiae Naturalis Principia Mathematica*, or *Mathematical Principles of Natural Philosophy*. While historically correct, this answer is seriously misleading. Newton's *Principia* is the single work that most effected the divorce of physics, and hence of science generally, from philosophy. [...] Correspondingly, to say that Newton's *Principia* is a work in philosophy is to use this term in a way that it rendered obsolete.[11]

I identify with large parts of Grant's, and Cohen and Smith's analyses of the relation between philosophy and science in Newton's work. Grant notes accurately that whatever title Newton chose to call his work, *Principia* is a work of science. In *Principia*, propositions concerning laws of nature make an axiomatic system, which includes mathematical proving, and contesting these proofs together with astronomical observations and pendulum experimentation. Newton's *Principia* is pronouncedly physics. No verbal trickery can falsify this point. Cohen and Smith acutely point out that, judged from the viewpoint of intellectual history, *Principia* was a truly revolutionary work as it started to separate the domains of philosophy and physics. It seriously questioned the authority of metaphysics as the first philosophy. Cohen and Smith (2002: 2) rightly indicate that after the publication and the reception of the *Principia*, questions of "what physically exist would no longer fall within the scope of traditional metaphysics."

All the same, I do not think that the analyses provided by Grant, and Cohen and Smith on the relations of philosophy and physics justify their stark conclusions. I argue that there are two reasons for this, and I shall propose a more cautious conclusion. First, their analyses give us too narrow a description of what the early modern natural philosophy was about. It is more diverse in its content and goals than they led us to understand. To appreciate this diversity, we have to examine closely the meaning of the terms "science" and "philosophy," and, more importantly, their scopes. Second, the analyses by Grant, and Cohen and Smith assume a dichotomy—not just a marked difference, as I assume— between philosophy and physics. I do not think this dichotomy is tenable. If we define natural philosophy as a gray area between philosophy and physics, as I have done in the previous section of this book, I think we can find such natural philosophy in Newton. In addition to mathematical physics, he was also involved in epistemic, ontic, and metaphysical analyses; that is, in philosophy. Accordingly, it is best to phrase Newton's contribution in terms of natural philosophy, a term that was also given the privilege to be in the subtitle of his magnum opus.

On the Early Modern Notions of Science and Philosophy

How was the word "science" defined in its early modern context? Although this word corresponds in part to our contemporary understanding of what science is, the match is by no means perfect. In many circumstances, the word "science," together with its Latin, French, and Italian cognates, was used to denote bluntly "knowledge." An identifying characteristic of knowledge was thought to be certainty. This is evident, for example, in Descartes' epistemology. He defines *scientia*, that is, knowledge, in terms of doubt. For example, we know that the foundational *cogito* argument is certainly true because we cannot doubt it (see Newman 2014). In addition to *cogito*, other typical exemplars of certainty were thought to be mathematical truths, including the theorems of geometry, arithmetic, and algebra. This is different from our contemporary understanding of science. I would imagine that today many scientists and philosophers of science would say that science is fallible, and that it does not produce impeccable certainty. Also, theology and metaphysics were sometimes used synonymously with science (Hatfield 1996: 495), but now we would not equate them with science.

In the early modern parlance, the term "philosophy" is actually closer to the meaning that we would ascribe to "science." Chambers' dictionary (1728: 803) defines "philosophy" as "the Knowledge or Study of Nature and Morality, founded on Reason and Experience." It mentions that "the Word Philosophy is used in various Significations among ancient and modern Writers," including, for example, the traditional meaning of the word philosopher, "a Friend or Lover of Wisdom." Chambers also introduces a distinction between natural and moral philosophies. Hence in this context "philosophy" is identified with special sciences. Natural philosophy is composed of, among others, physics, astronomy, and anatomy. Moral philosophy, for its part, consists of disciplines like psychology, history, economics, sociology, or their proto-forms.

The scope of "natural philosophy" is wider than that of "natural science" or "physics." Many philosophers who investigated nature during the seventeenth and eighteenth centuries, such as Descartes, Boyle, and Leibniz, included aspects to their works that we would classify not only as specific scientific doctrines, but as philosophy and theology (Janiak 2015: 18–19). Newton's work is also broader in scope than natural science, or physics, as we understand it nowadays. To show this, it would be easy to point out the voluminous efforts he took with biblical exegesis and alchemy in his unpublished papers. In reading these papers, we immediately see that his interests were much wider than mathematics or empirical science. But as the work under scrutiny here is his *Principia* (comprising also its later editions from the years 1713 and 1726), I shall concentrate only on it. In addition to its scientific aspects, the book contains an ample amount of philosophical as well as theological dimensions. We can see this especially in the following themes treated by Newton: causal account of laws of motion, the concepts of absolute space and time, arguments in favor of his methodological commitments, and in his position on the applicability of mathematics. I will examine these points rather briefly, because I will later provide more extensive analyses on the topics of causal nature of laws, the concepts of space and time, methodological commitments, and the relation of mathematics to nature. Here the motivation of introducing these notions is to argue that the *Principia* contains gray areas in which philosophy and physics overlap, thus satisfying the definition of natural philosophy as developed in this chapter.

In Newton's own words, as expressed right at the beginning of his *Principia*, the primary concern "of philosophy seems to be to discover the forces of nature from the phenomena of motions" (Newton 1999: 382). The background of the *Principia* can be, in part, traced back to the late medieval Aristotelian tradition of natural philosophy.[12] A quintessential point of this philosophy, which Newton adopts, is that nature is organized causally. Causes are ontologically prior to their effects but the effects are epistemically prior to their causes (Ducheyne 2012: chapter 1).

Newton's dynamics is based on the idea that force is the true cause of change of motion. Forces are treated as causes, and accelerations are treated as their effects. Forces are not directly observable. We do not, for example, see the gravity of the earth pulling the moon and thus keeping it in its orbit. Our knowledge of causes is founded on the observable and measurable changes in the motion of objects. To illustrate this point, consider the following example: I throw a rock. Newton's first law says that the object should move in a straight line to infinity. But this is not what we shall perceive. The trajectory of the rock will be close to a parabola as the object is affected by the earth's gravity. The rock will, due to air resistance and friction, eventually come to a halt. Newton infers that we can know the observed effects, the changes in the motion of the rock, because there is a common cause to it, namely gravity.

In the *Principia*, gravity is a proximate cause of acceleration. Gravity has its proximate cause but Newton eschews hypotheses concerning it (or at least he does not support a specific hypothesis of gravity's cause[13]). There is a finite number of proximate causes in nature. The most remote, ultimate cause is theistic God. Newton comments on the origin of the universe (our solar system) in the General Scholium as follows: "This most elegant system of the sun, planets, and comets could not have arisen without the design and dominion of an intelligent and powerful being."

Newton is explicit that there is a categorical difference between absolute and relative motions:

> [A]bsolute and relative rest and motion are distinguished from each other by their properties, causes, and effects [...] The causes which distinguish true motions from relative motions are the forces impressed upon bodies to generate motion. True motion is neither generated nor changed except by forces impressed upon the moving body itself, but relative motion can be generated and changed without the impression of forces upon this body.
>
> *Principia*, first Book, Scholium of the Definitions

The concept of "absolute motion," to wit, acceleration, is not meaningful without a fixed point of reference that is truly at rest. Motion is a comparative phenomenon; if we denote something as moving, there needs to be something at rest with respect to the motion. As Newton treats acceleration as an absolute quantity, it is not sufficient to refer to the observable surroundings of a moving object. Otherwise all motion would be relative, and the objective reality of his laws of dynamics would become questionable. Therefore, the quantities of space, time, and motion cannot be "conceived solely with reference to the objects of sense perception," as he notes in the Scholium of the Definitions.

Newton assumes that causal laws of motion require absolute space and time. Absolute, entirely homogenous, three-dimensional Euclidian space exists in its own right, independent of anything but itself and God. Likewise, time, what Newton also calls duration, is absolute (and universal). Objects impressed by a continuous force move at fixed spatial intervals in an absolutely uneven amount of time, whereas inertially moving objects move at fixed spatial intervals in an absolutely even amount of time.

Newton's reliance on the relation of cause and effect and his argument for absolute space and time designate that the *Principia* has important philosophical dimensions. The ambition of Newton's philosophy is to show that his dynamic laws are objective descriptions of the material world. He is an ontological realist with regard to the force of gravity[14] (and he takes other unobservables into his ontology, like mass and absolute space and time). This ambition cannot be realized without philosophy, without holding a position on what exists and how we come to know it. The *Principia* is not a philosophically neutral work.

Other related philosophical aspects of Newton's work are evident in the section of Rules for the Study of Natural Philosophy. It is replete with answers to philosophical problems of induction, explanation, hypotheses, and the reliability and generalizability of results. It also contains argumentation for the natural-philosophical methodology. What is more, Newton argues for a specific philosophy of mathematics. He believes that geometry concerns physical space. We see this at the very beginning of the *Principia*: "geometry is founded on mechanical practice and is nothing other than that part of *universal mechanics* which reduces the art of measuring to exact propositions and demonstrations." This is a very traditional position in the philosophy of mathematics on the relation between mathematics and nature. For example, Archimedes (1879) and Euclid (1945) applied geometry in their physical investigations. Demonstrations concerning torque, buoyancy, or optical ratios were thought to concern physical reality directly. This is not a trivial assumption. We might think that, for example, mathematics is *a priori* and analytic, and does not in itself concern reality. My point here is not to decide which of the two philosophies of mathematics is right. The point is that Newton has a philosophy of mathematics.

To characterize Newton's major work as a scientific work of physics is approximately true. But such a characterization assumes an all-encompassing dichotomy between physics and philosophy. This dichotomy is fallacious. In the *Principia*, we find a gray area in which both philosophy and physics operate. Based on the idea of this gray area, which I understand to be necessary for defining natural philosophy, it is more accurate to say that the *Principia* is a piece of natural philosophy, which stresses the mathematical and empirical treatment of natural phenomena, and in which theology also has a critical role to play.[15] To say that Newton was doing exclusively mathematical physics is simplistic because, to use our contemporary constructs, he was also engaged in metaphysics, epistemology, philosophy of science, philosophy of mathematics, and theology. His positions on these matters are not trivial ramifications of his scientific research, but products of various interrelated philosophical positions.

In discussing the relation between philosophy and physics, it is also worth pointing out that during revolutionary periods in physics, for which Newton's *Principia* stands out as a clear example, philosophy often becomes involved with the formulation of new theoretical frameworks. Here we may draw on Thomas

Kuhn's notion of a paradigm. In the normal phase of scientific research, scientists are focused on problem solving. After the emergence of enough anomalies that challenge and ultimately revoke the received paradigm, scientists need to take philosophical puzzles very seriously. It is of no accident that Newton—and later the forerunners of relativity and quantum mechanics—assessed traditional philosophical problems (Kuhn 1996: 88). Although Newton's education involved late medieval Aristotelian natural philosophy, as well as contemporaneous Cartesian natural philosophy (Smith 2007: section 1.2), his work was transformational in many respects. To make convincing arguments for his peers, Newton was obliged to detail his methodology and several philosophical positions and presuppositions that were relevant in estimating the core results of the *Principia*. The famous debate he had with Leibniz, in which Clarke was also the spokesperson for Newton, did not concern the particulars of astronomical data, and not even the mathematics of the inverse-square law. Instead, the quarrel was about constitutive philosophical issues, among others, the metaphysics of forces, the proper notion of causation, and the constraining character of the principles of sufficient reason and intelligibility in the acquisition of knowledge (see Janiak 2015: 24–6). To take part in such a debate is essentially doing philosophy; the debate is a form of dialogic argumentation that centers on what there is and how we come to know it.

I trust my treatment of natural philosophy in this chapter has provided a cogent definition of it. In the next chapter, I shall turn my attention to Hume's relation to natural philosophy. The outcome of the chapter is that while Hume is first and foremost a moral philosopher, his work is also engaged in natural philosophy.

Science of Human Nature and Natural Philosophy

This chapter deals with the relation between Hume's science of human nature and natural philosophy. I delve into Hume's science of humanity by focusing on the theory of ideas. I then delineate his relation to natural philosophy. I outline Hume's education and analyze the topics of natural philosophy that he touches upon. I argue—and agree with the traditional view—that Hume's primary concern is the science of human nature. I also show how his work is in part engaged with natural philosophy.

Hume's Ambition: Science of Human Nature

In his *Treatise of Human Nature*, Hume's primary objective is to establish a new science of human nature. His goal is to become "thoroughly acquainted with the extent and force of human understanding." This goal can be reached if we "explain the nature of the ideas we employ" and "the operations we perform in our reasonings." Such an explanation is "the sole end of" Hume's "logic" (T Intro 4; SBN xv). In Miren Boehm's (2013a: 58–9) interpretation, "logic" for Hume means the following:

- "General study of the nature of our ideas and of the principles and operations involved in reasoning."
- Tracing of "*causal* connections between the elements of the mind or perceptions."
- Explanation of the "operations of the mind in terms of principles of *association* of ideas."
- Description of "how the mind works, including how it reasons causally in terms of how the mind *naturally* led to consider one idea and then another."

The first element of Hume's logic that Boehm identifies, the "study of the nature of our ideas," is a very typical objective for early modern philosophy.

Here Hume relies on Descartes' theory of ideas. For his part, Descartes borrowed the idea theory, albeit in a modified form, from Plato. In Descartes, philosophical inquiry is essentially an investigation of mind's ideas (Newman 2014: section 1.2). Hume's version of the theory of ideas is of course markedly different from Descartes', because Hume rejects innatism. His starting point is analogous to Locke's. The mind begins to have ideas when it begins to perceive. To ask when a human being first acquires ideas, "is to ask," Locke writes in his *Essay Concerning Human Understanding* (2.1.9), "when he begins to perceive, having ideas and perception being the same thing."

According to Hume's copy principle, perception is divided into two: ideas and impressions. The difference of the two is a difference of degree, not type. Impressions are more forceful and lively than ideas. All simple ideas resemble some simple impressions. Ideas are copies of impressions. Following Locke, Hume thinks there are two sorts of impressions: sensations and reflections.[1] Sensations are epistemically and ontologically primary to reflections. The former are the original impressions that strike the mind. Impressions of sensation arise in the mind "originally, from unknown causes" (T 1.1.2.1; SBN 7). Impressions of reflections are produced by ideas already acquired from impressions. For example, eating tasty food gives me an impression of pleasure, and hitting my toe on the wall causes a feeling of pain. When the impressions of pleasure or pain return to the mind, the mind causes new impressions of reflection of "desire and aversion, hope and fear" (T Sections 1 and 2; SBN 1–8).

Nothing can be in the mind if there is no simple impression causing its idea in the first place (T 1.1.1.1–7; SBN 1–4, EHU 2; SBN 17–22). An exception to this is the missing shade of blue.[2] Say that an adult, who has previously in her life perceived a diverse spectrum of colors, is presented a color chart made of shades of blue. She perceives the contiguous shades successively, but at one instant, in the middle of the show, a particular shade is missing, and she sees a blank. Hume believes that this person can, from her own imagination, perceive the shade. This serves "as a proof, that the simple ideas are not always, in every instance, derived from the correspondent impressions" (T 1.1.1.10; SBN 6). This is still a minor exception to the (causal) copy principle, so it should not alter its status as the general maxim of Hume's philosophy. The principle is still the starting point for Hume's science of human nature in his *Treatise* (1.1.1.7; SBN 4): "The *full* examination of this question is the subject of the present treatise; and therefore we shall here content ourselves with establishing one general proposition, *That all our simple ideas in their first appearance are deriv'd from simple impressions, which are correspondent to them, and which they exactly represent.*" Hume's goal, as he also puts it in his first *Enquiry*, is to provide a "mental geography" by mapping the cognitive structures of the human mind. He seems to be happy in going "no farther than this mental geography, or delineation of the distinct parts and powers of the mind" (EHU 1.13; 13).

Regarding reasoning, Hume identifies two different types: demonstrative and probable. I will analyze Hume's account of reasoning thoroughly later in Chapter 5 of this book. For this chapter, it is enough to lay down the basics. The first type of reasoning includes relations of ideas, the latter the relation of causation. In demonstrative reasoning, we proceed from an idea to its adjacent idea by means of intuition. A sequence of intuitions counts as a demonstration. Probable reasoning is causal. Causation is identified with experience. By experience, a notion that will be explicated in Chapter 3 of this book, Hume means observation and memory of two species of objects or events being constantly conjoined (T 1.3.6.2; SBN 87). All factual reasoning is founded on the relation of cause and effect. Therefore reasoning concerning any fact, whether human or natural, is founded on experience (EHU 4.14; SBN 32). The way we acquire factual knowledge is by perception and experience.

The crux of Hume's philosophy is to explain the way we get ideas and how we reason with them. Boehm calls this Hume's foundational project. All sciences, or branches of learning, are to a considerable extent dependent on the science of human nature. They depend on Hume's logic. This is "because to do science, scientists must *employ ideas and engage in reasoning*," Boehm (2013a: 59) argues. Thus Hume himself puts it as follows:

> 'Tis evident, that all the sciences have a relation, greater or less, to human nature; and that however wide any of them may seem to run from it, they still return back by one passage or another. Even *Mathematics, Natural Philosophy, and Natural Religion*, are in some measure dependent on the science of Man; since they lie under the cognizance of men, and are judged of by their powers and faculties. 'Tis impossible to tell what changes and improvements we might make in these sciences were we thoroughly acquainted with the extent and force of human understanding, and cou'd explain the nature of the ideas we employ, and of the operations we perform in our reasonings.
>
> T Intro 4; SBN xv

One reading of this quote is that Hume privileges science of human nature over natural philosophy. Boehm argues that natural philosophy, like Newton's, is subordinate to Hume's science of humanity. She writes:

> Hume conceives of his own science of man neither as subordinate nor even as standing alongside Newton's natural philosophy. In fact, Hume envisions his

science of man as occupying a *foundational* role with respect to all sciences, *including* Newton's (T Intro 8). Hume understands the natural philosophy of Newton to be *dependent* on his science of man (T Intro 4–5).

Boehm, 2016: 1–2

To use our contemporary constructs, we can characterize Hume's focus as being first and foremost in cognitive science or cognitive psychology, even in phenomenology, or their proto-forms.[3] His primary interest is not physics, or some other aspect of natural philosophy. His contributions to "philosophy and general learning" center around the study of the human mind. Even though Hume probably wanted to imitate the success of experimental natural philosophy (of Newton, for example) in his moral philosophy, his main concern is in mapping the cognitive structures of the mind (Harris 2015: 85).

In Hume's explicit statement, he tells the reader that he does not explain the "natural and physical causes" of our perceptions, because this task is for "the sciences of anatomy and natural philosophy" (T 2.1.1.2; SBN 275–6). To explain this point, we may use colors as an example. Hume's humanistic science is interested in color perception (see Kervick 2018). He explains the difference between perceiving specific colors by referring to different discrete impressions that create the ideas of these colors. This is as far as Hume's explanatory science of humanity goes. He does not intend to explain the difference by positing lengths of light rays or frequencies of light waves. Such theory about the physical composition of light is necessary for explaining the spectrum of light from the perspective of physics. Hume does not subscribe to such unobservable posits, but contents himself with explaining the perceivable spectrum via discrete simple impressions.[4] Hume's ambition is to establish a human science which is different from any of the natural sciences, like physics.

This leads us to the question: What is the relation of Hume's science of human nature to natural philosophy? For the most part, Hume is doing moral philosophy. I will still argue that on many occasions his work touches upon themes relevant to natural philosophy. Hume was rather well-versed in the physics of his time, which gave him the means to articulate many interesting natural philosophical positions.[5]

Hume's Education in Natural Philosophy

Hume is not placed in the canonical listings of early modern natural philosophers like Newton, Boyle, Huygens, or von Linné. To his contemporaries, Hume was known as a historian and essayist. He was not perceived as a natural philosopher, or, as we might roughly put it, a natural scientist. He did not have a similar kind of competence in mathematics and physical science like Leibniz or du Châtelet did. Before turning to Hume's engagement in natural philosophy, it is important to first scrutinize his education in the field. To understand Hume's education in natural philosophy, and the broader scientific culture of his time, there is no better scholarly article on the matter than Michael Barfoot's (1990) "Hume and the Culture of Science in the Early Eighteenth Century." Therefore, concerning Hume's education, I shall draw heavily on Barfoot's treatment of the issue. Another central source for understanding Hume's education is Margaret Schabas' (2005) book, *The Natural Origins of Economics*.

During the 1724–1725 session, Hume attended Robert Steuart's natural philosophy class at the University of Edinburgh. Hume's possible lecture notes from the class have not survived. However, an account of the syllabus of Steuart's class was published in *The Scots Magazine* of 1741 (371–2). It states the following:

> He teaches, first, Dr John Keill's *Introductio ad veram Physicam*, and the Mechanics from several other authors. After that he teaches Hydrostatics and Pneumatics from a manuscript of his own writing. Then he teaches Dr David Gregory's *Optics*, with Sir Isaac Newton of *Colours*; describes the several parts of the Eye, and their uses, with the phenomena of Vision, and describes all the different kinds of microscopes and telescopes. Then he teaches Astronomy, from Dr David Gregory's *Astronomy*, with some propositions of Sir Isaac Newton's *Principia*, and the Astronomical observations both ancient and modern. He likewise shews a set of Experiments, Mechanical, Hydrostatical, Pneumatical and Optical. He expects that before any student enter his class, he has read Geometry, &c. with Mr MacLaurin one year at least.

The account of the syllabus suggests that the class was Newtonian in spirit, as Gregory admired Newton, Keill was known as Newton's disciple, and MacLaurin was a close affiliate of Newton. However, Barfoot infers that the class was also devised by using Jacques Rohault's Cartesian textbook *System of Natural Philosophy*. This suggests that Hume's education involved both Cartesian and Newtonian elements. This is not surprising given Steuart's reputation; he has been characterized as being a Cartesian early in his career, but then his concept of natural philosophy took a shift toward Newtonianism. It is also important to note that the vague doctrinal labels "Newtonianism" and "Cartesianism" (or its close affiliates of "Experimental" and "Speculative") are by no means exhaustive. The course also included the study of Boyle, whose natural philosophy emphasizes experimentation but also leans on a mechanistic account of corpuscles.[6]

There is circumstantial evidence that Steuart's course centered around experimental aspects of natural philosophy, hydrostatics, and pneumatics in particular.[7] It is probable that Hume became acquainted with experimental demonstrations. It is also probable that he had quite a good understanding of the mathematics of his time, as this was necessary for following Steuart's class. Most students at the University of Edinburgh received a basic, philosophically oriented training in mathematics that privileged geometry over algebra.[8] The more advanced students were expected to have a command of fluxions, so that they could read Newton's *Principia* in some detail. According to Barfoot, there is circumstantial evidence that Hume belongs to the category of the more advanced students. Hume's lecture notes from his extramural mathematics instructor George Campbell reveal that Hume knew the basics of fluxions. His notes do not reveal how well-versed he was in the application of calculus in physics. However, Campbell's teaching did concentrate on the lemmas of the first Book of the *Principia*, section 1.

There is more circumstantial evidence that Hume was no stranger to mathematics. He treats Euclidian geometry at length in the first Book of the *Treatise*. We now know that among his unpublished manuscripts, there were two essays on mathematics; one was written by Robert Wallace, the other might have been authored by Hume himself.[9] The fact that Hume later became well acquainted with Maclaurin further denotes his affinity to mathematics (Schabas 2005: 65). Hume also served as a secretary to the Philosophical Society of Edinburgh. He edited scientific papers on physical sciences such as astronomy, electricity, and optics (and on life and medical sciences) (Sapadin 1997: 338).

Hume's Interest in Various Natural Philosophical Issues

The previous section established that Hume was competent in the physics of his time. He was not brilliant like his contemporaries du Châtelet and Reid, but he had enough information to have an early modern philosophy of physical science. In this section, I wish to take a closer look at what topics Hume thought were interesting in natural philosophy, and how these different topics are related.

We may start with Hume's concept of a law of nature. This is evidently a notion used in natural philosophy.[10] Hume is concerned with articulating what laws of nature are, so his work definitely engages in natural philosophy. He does not stay merely in the realm of the science of human nature.

Hume's conception of laws is related to his conception of causation, as laws of nature are matters of fact that are founded on the relation of causation. The knowledge we have on the relation of causation is acquired by experience, or, in Hume's language, also by experiments. Both of these conceptions, of laws of nature and of causation, are related to his understanding of the reality of forces. Hume's position on the ontology of forces cannot be understood without his copy principle. In turn, the copy principle is the foundation of his philosophy of space and time, and it is of central importance for scrutinizing Hume's skepticism concerning the existence of a vacuum. Moreover, propositions concerning laws of nature can be expressed in mathematical terms, so to understand Hume's conception of laws, it is necessary to investigate his conception of propositions. As Hume draws a sharp distinction between the propositions concerning relations of ideas and matters of fact with his "fork," how should we then understand the epistemic status of propositions of mixed mathematics, such as the propositions concerning the law of conservation of momentum, or the laws of Newtonian dynamics? As the propositions of pure mathematics are absolutely necessary, what about the propositions of mixed (applied) mathematics? Do laws which can be expressed in mathematical terms instantiate necessity, or mere regularity? How does mathematics actually relate to nature? Is the book of nature written in the language of mathematics, or is mathematics just a useful tool for expressing the magnitudes of causes and effects in a precise and predictable manner? All these topics are tightly connected.

Hume did not work in an intellectual vacuum, so understanding his natural philosophy requires putting his arguments into a proper historical context. Many of Hume's viewpoints are intrinsically related to early modern natural philosophy. A variety of elements in his thoughts are reminiscent of Cartesian cosmology, such as the assimilation of space to extension, critical stance on the existence of a vacuum, and the assumption of mechanism in causal interactions of bodies. In some other cases, his positions are closer to a typically Newtonian natural philosophy: Hume is a critic of hypothesis and speculative philosophy, he has a notion of inductive proof, and he rejects the Leibnizian principles of intelligibility and sufficient reason. Boyle's experimentalist program incorporated many items that are of central importance to Hume: the category of matter of fact as the foundation of physical knowledge, and the requirement of witness testimonies in assessing conditional probabilities. Hume's position on the ontology of forces somewhat resembles Berkeley's views. The concept of force is a mathematical instrument that enables scientists and engineers to make predictions; but as forces are unobservable, their real nature is not known to us. Leibniz's distinction between "truths of reason" and "truths of fact" precedes Hume's fork. Leibniz's relationist ontology concerning space and time, according to which space and time are not entities by their own right, is very congenial to Hume's position. His application of the term "mixed mathematics" has a history: The term was applied by other early moderns such as Francis Bacon and Jean le Rond d'Alembert. In this generally Aristotelian tradition of thinking about mathematics, which strictly distinguishes between pure and applied mathematics, Hume's treatment of mixed mathematics leads to a salient problem about the epistemic status of applied mathematics and to the question of how mathematics relates to nature.

Is Hume Doing Natural Philosophy?

As I understand it, Hume's work engages in natural philosophy. This is not to say that Hume is first and foremost a natural philosopher. His philosophy centers around explaining the cognitive structure of the human mind. He is also very critical of some parts of natural philosophy. Boehm (2013a: 57) argues that Hume's logic adjudicates "on questions within natural philosophy." I am partly sympathetic to Boehm's interpretation. The problem is that it is one-sided. It would be a false dichotomy to claim that Hume *either* criticizes natural philosophy *or* supports it. Criticism and support of parts of natural philosophy are not mutually exclusive but inclusive.

A particularly good example of the falsity of this dichotomy can be found in Hume's treatment of the ideas of space and time. It is true that his logic confines the application of these ideas within natural philosophy. We do not have impression-based ideas of absolute space and time, since these abstract ideas would have to be caused by finite simple impressions. As we lack the ideas of absolute space and time, we cannot even think them. The words "absolute space" (roughly, pure extension without matter) and "absolute time" (roughly, an absolute flow that is independent of perceivable change) cannot be annexed to any impression-based ideas. Newton argued for these absolute structures to make a difference between absolute and relative motions. Although the argument is explanatorily important and very desirable, it contains invalid reasoning for Hume. He cannot subscribe to this crucial aspect of Newton's natural philosophy because the putative absolute space and time are imperceptible. Such structures, if they exist, fall beyond the scope of human understanding. Hume's copy principle does not include an inference from non-observability to non-existence, but given our limited cognitive capacities, we cannot know whether such entities exist or not.

It would be wrong to say that Hume's position on this issue is altogether free from the physics of his day. His understanding of space as extension, his critical stance on the vacuum, and his relationist ontology regarding space and time are very similar to Descartes' and Leibniz's natural philosophical positions[11] (I will deal with the similarities between Descartes and Hume later in Chapter 6). Hume's treatment of these issues is not, of course, entirely independent of his science of human nature, as he is interested in explaining the way we get (or do not get) the ideas of space and time, vacuum, and so on. But it would be equally false to say that Hume's science is an entirely independent branch of study that is by no means affected by the preceding history of natural philosophy. There is no need to assume a dichotomy here.

Consider another philosophical item that is important both for Humean humanistic sciences and physics: causation. According to Hume's methodology in the *Treatise* (1.3.2.4; SBN 74–5), he begins by examining the origin of our idea of causation. He notes that sensory qualities of objects do not cause the idea of cause. Hence the idea "of causation must be deriv'd from some *relation* among objects" (T 1.3.2.6; SBN 75). By this method, tracing the sensory origins of our ideas, Hume finds causes and effects to be contiguous and successive. Did Hume arrive at this result solely from the viewpoint of his science of human nature? No. Contiguity and succession are restrictions for causation in the Cartesian–mechanistic paradigm. We see one object moving, it touches the other, and causes it to move. By and large, such thinking about causation leans on Cartesian cosmology. The rules by which to judge causes and effects are very hospitable to mechanistic physics, but in tension with a dynamic one (T 1.3.15.1; SBN 173). Because Hume assumes and argues for one type of natural philosophy, he also restricts Newtonian dynamics, which does not include a reference to contiguity or succession. In this sense, Hume can be reasonably interpreted as contributing to both science of human nature and natural philosophy, although he does not present himself as doing the latter.

Further, it is not clear whether the science of human nature really is epistemically privileged over natural philosophy, as Boehm maintains.[12] For one thing, Hume is clear that natural philosophy temporally precedes his moral philosophy: "the application of experimental philosophy to moral subjects should come after that to natural at the distance of above a whole century" (T Intro 7; SBN xvi). This indicates that Hume has had to learn from experimentalism in natural philosophy. Importantly, consider also the subtitle of the *Treatise*: "Being an Attempt to Introduce the Experimental Method of Reasoning into Moral Subjects." Here Hume speaks about an "attempt to introduce" the experimental method from philosophy of nature to moral philosophy. The title of the *Treatise* in its entirety does not assert any privilege to the science of human nature: it asserts that the experimental method of natural philosophy should be taken as the method for the science of human nature. These passages, unlike those quoted by Boehm (T Intro 4–5; SBN xv, Abstract 3; SBN 646), do not indicate any hierarchy between natural and moral philosophies but a methodological continuum between the two.

There is even contrary evidence to Boehm's reading. Hume applies natural philosophy to parts of his human sciences. In explaining the association of ideas, he compares the gravitational attraction of bodies to mental attraction (T 1.1.4.6; SBN 12). Given Hume's idea theory, it is an open question as to how ideas in the mind do not just randomly fluctuate but are regularly ordered. In the third section of his first *Enquiry*, he offers three principles of association: resemblance, temporal and spatial contiguity, and causation. Perhaps imitating Newton's criticism of hypotheses, Hume does not intend to explain why there are these principles; rather, he is satisfied in establishing that there are principles according to which the mind works, as there are principles according to which physical stuff moves (Morris and Brown 2014: Section 4.3). So Hume implements Newton's natural philosophical methodology into his cognitive psychology, which is what his research is mainly about. Hume's psychology is not the only part of the science of human nature that draws on natural philosophy. Economics is also part of Humean science. Hume had studied hydrostatics, and adapted models from this branch of physics to his theory on the circulation and flow of money (Schabas 2001: 412). If the foundational project reading were correct, Hume's implementation of the models of the physical sciences to his humanistic science would not make sense.

Moreover, it is also questionable whether natural and human philosophies are altogether different. In the previous chapter I argued, from a systematic philosophical viewpoint, that natural philosophy is a gray area between physics and philosophy; an area that covers both, but which is not only about the other. Tamás Demeter's (2017) study on the intellectual background of Hume's science shows that Scottish enlightenment philosophers usually conceptualized natural and human phenomena in similar ways. The methodology, reasoning, and ideology concerning the two are strikingly similar. Demeter's (2017: 5) study reveals that: "Enlightenment philosophy in Scotland—and early modern philosophy in general—should be seen as an integrated enterprise of moral and natural philosophy and conceived as intellectual enterprises that developed hand in hand."

Demeter points out that why we see the two domains so differently today is because we are still under the influence of Snow's "two cultures" dichotomy. This dichotomy does not correspond to the early modern intellectual world. Scholars in that period pursued a unified explanation of the interrelations of the physical, physiological, mental, ethical, and theological aspects of the world (Demeter 2017: 13). The seventeenth- and eighteenth-century studies on the foundational issues of human knowledge and the nature of reality took place just before different specific disciplines started to emerge as separate domains with their own scopes of investigations.

In light of this background, we may advance a reading of Hume's foundationalism different from that of Boehm. To borrow from the formulation of Demeter (2017: 6), Hume also "understands his own project as foundational." But this does not mean that other disciplines are subordinate to Hume's project. Demeter clarifies the significance of the foundational project: it is "a critical work that we cannot dispense with before immersing ourselves in other cognitive enterprises." In the words of Paul Stanistreet (2002: 18), "we will better understand the scope and limitations of the natural sciences once we understand the limitations of the human powers and faculties we employ in them."[13] According to this view, Hume explores the conditions and limits of human knowledge. To return to the example of Newton's views of space and time, we do not know whether such unobservable structures exist because we lack their putative ideas. Lack of knowledge concerning natural philosophical posits does not entail the inferiority of natural philosophy in any way. Rather, it shows that investigations about the human and natural phenomena have the same restrictions. Both are experimental disciplines in which hypotheses and imagination have only a very limited role. Hume himself is clear on the convergence of the two disciplines. This is true especially in the way the two acquire knowledge. In his first *Enquiry* (8.7; SBN 83–4), this is what he says on the relation between the investigations of the natural and the moral phenomena:

> These records of wars, intrigues, factions, and revolutions, are so many collections of experiments, by which the politician or moral philosopher fixes the principles of his science, in the same manner as the physician or natural philosopher becomes acquainted with the nature of plants, minerals, and other external objects, by the experiments which he forms concerning them.

The above quote does not establish that all experiments are alike, and none of this implies that Hume is first and foremost a natural philosopher. To clarify this point, it is useful to compare his views concerning natural philosophy with his metaphysical positions. The motivation of this analysis is to show that Hume in part criticizes metaphysics *and* in part has a metaphysics, *in the same way* as he criticizes parts of natural philosophy *and* in part has a natural philosophy.

Comparing Metaphysics to Natural Philosophy

Hume has traditionally, perhaps due to the influence of Kant and logical positivism, been received as a critic of metaphysics. This traditional reception is partly true. Hume is critical of a considerable part of metaphysics, of "school metaphysics and divinity," as he says himself (EHU 12.34; SBN 165). For example, he rejects the following metaphysical enterprises:

- The Cartesian philosophy of mind based on the notion of substance in which perceptions inhere, as well as unified self and synchronic personal identity.
- Leibnizian principles of sufficient reason and intelligibility.
- The possibility of synthetic *a priori* knowledge. (Hume nowhere uses the term "synthetic *a priori*." Kant took Hume to have anticipated the denial of synthetic *a priori* judgments, which Kant himself thought crucial for the possibility of metaphysics.)

Below I will analyze these three critical points, and then move on to explain how some parts of Hume's philosophy are rooted in metaphysics.

Criticism of Cartesian Metaphysics

Minds and bodies are substances for Descartes.[14] Minds are discrete entities, in which attributes like thinking and modes like doubt, inhere. In the section "Of Immateriality of the Soul" of the *Treatise* (1.4.5.4; SBN 233), Hume asks: "those philosophers, who pretend that we have an idea of the substance of our minds, to point out the impression that produces it, and tell distinctly after what manner that impression operates, and from what object it is deriv'd. Is it an impression of sensation or of reflection?"

Hume does not deny the existence of substances. Perceptions are substances. They are discrete beings that are separable from one another, "and may be consider'd as separately existent, and may exist separately, and have no need of any thing else to support their existence" (ibid.). However, the Cartesian substance is for Hume "an unknown *something*, in which" the attributes and modes of the mind "are supposed to inhere" (T 1.1.6.2; SBN 16). There are no impressions of such putative being. Cartesian "substance" denotes a term without any impression-based idea annexed to it. When we examine "our idea of *substance*," and ask the critical question, "*from what impression that pretended idea is derived?*" we find that we are unable to produce an answer that satisfies the copy principle (Abstract 7; SBN 648–9). In the first *Enquiry* (2.9; SBN 21), Hume makes it clear that he uses his copy principle as a device to criticize metaphysics. "If a proper use were made of" the principle, we could "banish all that jargon, which has so long taken possession of metaphysical reasonings, and drawn disgrace upon them." William Edward Morris (2009: 441) calls this "his resolutely anti-metaphysical stance."

Criticism of Leibnizian Metaphysics

In his correspondence with Clarke, Leibniz formulates his principle of sufficient reason. In the second letter (1989b: 321), he characterizes the "*principle of a sufficient reason*, namely, that nothing happens without a reason why it should be so rather than otherwise." Donald Rutherford (1992) argues that Leibniz advances an even more restrictive principle than that of sufficient reason: the principle of intelligibility. To quote from Rutherford (1992: 35), Leibniz's commitment to his intelligibility principle means "that nothing happens for which it is impossible to give a *natural* reason, i.e., a reason drawn from the natures of the beings that belong to this world." This principle is echoed by Leibniz's (1996: 66) own words: "Whenever we find some quality in a subject, we ought to believe that if we understood the nature of both the subject and the quality, we would conceive how the quality could arise from it." There must be some reason why a given property is predicable of a subject in the natural world (miracles excluded). Such predication is conceivable because of the nature of the being in question. Leibniz's assumption is that it should always be possible to understand how a given quality arises from the nature of a given subject. Accordingly, nature is intrinsically and ultimately intelligible.

Hume eschews Leibniz's principles. He makes this clear by invoking Adam, the ideal but *a priori* epistemic agent (EHU 4.6; SBN 27, Abstract 11–4; SBN 650–2). Adam is equipped with a perfect sensory system and reason. Hume asks, can Adam predicate any properties of objects? Could he know, based on his immediate clear perceptions and perfect reasoning capacities, the attributes of water, flame, gunpowder, and lodestone? Does he know that water is potentially drowning to non-aquatic beings, that flame is hot, powder explosive, and that lodestone is attractive to metals? Hume's answer is an abrupt no, because there is no reason why any property is an attribute of any object. Even after an indefinitely large amount of experimentation, we could not provide a "satisfactory reason" for any such predications (EHU 12.25; 162). To know about the qualities of objects, we need causal reasoning. In turn, causal reasoning is based on experience, not on any *a priori* reasons. In the *Treatise* (1.3.1.1; SBN 69–70), Hume formulates his position as follows:

> 'tis evident *cause* and *effect* are relations, of which we receive information from experience, and not from any abstract reasoning or reflection. There is no single phaenomenon, even the most simple, which can be accounted for from the qualities of the objects, as they appear to us; or which we cou'd foresee without the help of our memory and experience.

Hume maintains that we can ascribe qualities to objects because the qualities are constantly conjoined with the objects. Predication is a matter of identifying regular patterns, not a matter of reason or intelligibility of nature, as Leibniz has it. Hume's empiricism is in stark contrast to a metaphysical ambition of discovering the primary qualities of matter or ultimate level of explanations.

Criticism of Kantian Metaphysics

This rubric is doubtlessly anachronistic. It can still be argued that Hume anticipates a criticism of Kantian metaphysics, namely the denial of synthetic *a priori* knowledge. Kant himself recognized Hume's critical take on metaphysics in the Introduction to his first *Critique* (KdRV, B 19–20):

> **How are synthetic *a priori* judgements possible?** [...] On the solution of this problem, or on a satisfactory proof that the possibility that it demands to have explained does not in fact exist at all, metaphysics now stands or falls. David Hume, who among all philosophers came closest to this problem, still did not conceive of it anywhere near determinately enough and in its universality, but rather stopped with the synthetic proposition of the connection of the effect with its cause (*Principium causalitatis*), believing himself to have brought out that such an *a priori* proposition is entirely impossible, and according to his inferences everything that we call metaphysics would come down to a mere delusion of an alleged insight of reason into that which has in fact merely been borrowed from experience and from habit has taken on the appearance of necessity.

Hume nowhere uses the notions of analytic and synthetic.[15] However, he denies the Kantian notion of synthetic *a priori* in two senses: 1) he claims that pure mathematics is the only demonstrable science, and 2) he disagrees with Kant in that the necessity and certainty that is typical to mathematics could be extended to concern factual-causal propositions. Points 1 and 2 are evident in Hume's words at the end of his first *Enquiry* (12.27; SBN 163)[16]:

> It seems to me, that the only objects of the abstract sciences or of demonstration are quantity and number [1], and that all attempts to extend this more perfect species of knowledge beyond these bounds are mere sophistry and illusion [2].

The upshot of Hume's criticism is that there is no special metaphysical knowledge. If a proposition is *a priori*, then it does not convey any information about the world, but solely about the relations of ideas. If a proposition is informative about the world, then it should be supported by experience and inductive-probabilistic reasoning.

Hume's Metaphysics

The previous cases show that Hume is a harsh critic of many aspects of metaphysics. His outright denial of substance dualism, principles of reason and intelligibility, as well as (the anticipation of the denial of) synthetic *a priori* knowledge question even the very possibility of Cartesian, Leibnizian, or Kantian metaphysics. As he famously ends his first *Enquiry* (EHU 12.34; SBN 165), works of metaphysics that do not include any mathematical or experimental reasoning should be scorched:

> When we run over libraries, persuaded of these principles, what havoc must we make? If we take in our hand any volume; of divinity or school metaphysics, for instance; let us ask, *Does it contain any abstract reasoning concerning quantity or number?* No. *Does it contain any experimental reasoning concerning matter of fact and existence?* No. Commit it then to the flames: For it can contain nothing but sophistry and illusion.

Some read this as indicating Hume's rejection of metaphysics. For example, A. J. Ayer (2001: 21) claims: "Of Hume we may say not merely that he was not in practice a metaphysician, but that he explicitly rejected metaphysics." Ayer quotes the concluding paragraph of the first *Enquiry* as evidence for his position. Then he notes that "Hume does not, so far as I know, actually put forward any view concerning the nature of philosophical propositions themselves, but those of his works which are commonly accounted philosophical are, apart from certain passages which deal with questions of psychology, works of analysis" (ibid.: 22).

Morris agrees with Ayer's positivistic reading. According to Morris (2009: 442), Hume "advocates the elimination of metaphysics and its replacement with an empirical, descriptive 'science of human nature' based on 'observation and experience.'"

These views can be challenged by showing that Hume has a nominalist background of metaphysics.[17] He contends that "every thing in nature is individual" (T 1.1.7.6). This is obviously a metaphysical position, very much like Descartes' dualism or Spinoza's monism. Donald L. M. Baxter (2007: 6) argues that Hume supports several metaphysical doctrines. Hume thinks, among others, that "only particular things exist," "alteration is contrary to identity," "the conceivable is possible," and that "there is no middle way between existing and not existing." None of these views is metaphysically neutral; Baxter even goes so far as to say that "Hume is a great metaphysician."

I think both the positivist and the superfluously metaphysical readings of Hume are problematic. It cannot be, as Ayer (2001) and Morris (2009) claim, that Hume eliminates metaphysics altogether. His copy principle is not metaphysically neutral as his version of the theory of ideas is imbued with nominalist metaphysics. Also, the final paragraph of the first *Enquiry* contains metaphysical assumptions. The critique of school metaphysics is based on his fork. In turn, his fork is founded on the doctrine of relations in *Treatise* 1.3.1. Hume provides what is in his mind a *complete* classification of modalities. An attempt to explain all modal categories is evidently a metaphysical enterprise. Moreover, if Hume would like to eliminate metaphysics, he would encounter similar problems as the positivists. Famously, the principle of verifiability turned out to be self-contradictory: it is itself neither analytic (math, logic, conventional truths) nor synthetic (empirical fact). Why is Hume's conclusion not self-stultifying? Why does Hume not throw his own book into the flames? Because his point is that works which do not include *any* of the two propositions are hogwash. Hume has no need to present himself as getting along without *any* metaphysics.

However, calling Hume a "great metaphysician" is also problematic. It downplays the stringent criticism Hume develops against several branches of metaphysics. In this regard, I think Morris (2009: 453) has it right: "A central part of his program is the profoundly anti-metaphysical aim of abandoning the *a priori* search for theoretical explanations that supposedly give us insight into the ultimate nature of reality."[18] I will further advance this point in the next chapter, which deals with experimentalism and critique of hypotheses.

The reasonable interpretation of Hume on metaphysics is that he both argues for and assumes some type of metaphysics, *and* criticizes some type of metaphysics. As in the case of Hume's criticism and his support of parts of natural philosophy, it would be a false dichotomy to say that Hume is *either* a destructive critic of metaphysics *or* a full-blown metaphysician. Some of Hume's philosophical positions pertain to metaphysics in the same way as some of his positions pertain to natural philosophy. His philosophy is not independent of metaphysical assumptions. Relatedly, none of Hume's views about experimentalism, causation, laws of nature, the ontology of forces, the nature of matter, the status of mathematics in science, and space and time is detached from natural philosophy. And it is the task of the forthcoming chapters to figure out how Hume engages with such natural philosophical issues.

3

Experimentalism

The post-Kantian history of philosophy (the twentieth-century history of philosophy in particular) divided early modern philosophy into two rival camps: the rationalists and the empiricists. According to this traditional narrative, Kant then synthetizes the two camps with his transcendental idealism. In recent years, the Otago school[1] has questioned the narrative. Alberto Vanzo (2016: 254) lists the shortcomings of the rationalist/empiricist dichotomy.[2] The dichotomy:

- pays too much attention to epistemological issues;
- underestimates important debates in other areas, including natural philosophy;
- mistakes, in some circumstances, empiricists for rationalists, and rationalists for empiricists;
- and, finally, makes arbitrary distinctions to hide common viewpoints between the empiricists and rationalists, like Hume's (the empiricist) affinity for Malebranche (the rationalist) on causation and for Leibniz (the rationalist) on space and time.

Instead of upholding the traditional narrative, the Otago school proposes that it should be modified with actors' categories. For this purpose, experimental and speculative philosophies are better labels. Vanzo (2016) and Peter Anstey (2012: 500) provide textual evidence for this claim by referring to bookseller John Dunton's student manual from the year 1692. In the manual, Dunton (1692) first divides philosophy into natural and moral, and then further subdivides the former into speculative and experimental. The divide is explicated by John Sergeant in his *Method to Science* in 1696:

> The METHODS which I pitch upon to examine, shall be of two sorts, viz. that of Speculative, and that of Experimental Philosophers; The Former of which pretend to proceed by Reason and Principles; the Later by Induction; and both of them aim at advancing Science.

Before the mid-seventeenth century, natural philosophy had been largely based on speculation. It was not seen as guiding action or devising artefacts. Knowledge was thought to be deducible with demonstrative inference from first principles. The traditional natural philosophers mentioned experiments, but in doing so they referred to previously established conclusions. The conclusions were derived from authorial textual sources and thought experiments, not from actual repeatable experiments or first-hand experience (Anstey and Vanzo 2016: 88).

In this chapter, I place Hume in the context of the British tradition of experimentalist natural philosophy. By applying the experimental/speculative divide, we can avoid focusing solely on epistemic issues,[3] and highlight the importance of the history of science (for this book, the history of physics in particular). This is not to say that I regard "empiricism" as a wrongful label in Hume's case. He is both a concept empiricist and an epistemic empiricist. His concept empiricism can be characterized as follows. He maintains that the origins of all our simple ideas are in impressions. All meaningful terms can be annexed to ideas, and these ideas can be broken into simple ones, which have correspondent simple impressions. Although imagination creates complex ideas, which themselves do not have sensory correlates, the component simple ideas have their origin in simple sensory impressions. In turn, Hume's epistemic empiricism can be characterized as follows. There is no *a priori* knowledge of reality. All factual propositions—roughly, all non-tautologous[4] propositions excluding relations of ideas—are justified with experience. We do not know what the world is like solely by thinking; this requires experience.

Regarding the experimentalist tradition, I limit my study to Boyle and Newton, and their relationship to Hume. I wish to show how the experimentalist positions paved the way for his epistemic empiricism. Hume borrows the notion of "experiment" from the natural philosophical tradition and replaces it with his notion of "experience."

Boyle on Probability and the Category of Fact

Before the mid-seventeenth century, "knowledge" and its corresponding term "science" were clearly distinguished from "opinion." Before Boyle introduced his new experimental method of inquiry in pneumatics, natural philosophers imitated the demonstrative and axiomatic sciences, like geometry. This was thought to produce impeccable certainty and universal assent (Shapin and

Schaffer 1985: 23–4). Only this would count as scientific knowledge. According to this largely Aristotelian ideal, an explanation is a form of a logical argument. The conclusion of a syllogism denotes scientific knowledge. To ascertain the conclusion, the premises of the argument need to be free from all error. Premises might be conclusions of previously proven arguments. As this kind of deduction cannot go on to infinity, there needs to be some basic principle, an axiom which itself is indubitable. It is essential that scientific arguments are based on such axiomatic premises, so that the conclusions of the arguments produce universal assent and certainty (DeWitt 2010: 51–4).

In contrast to the Aristotelian tradition, the British experimentalist tradition from the 1650s onward proposed that physical knowledge can yield probability. This breaks down the twofold distinction between knowledge and opinion. According to the experimentalist position, Steven Shapin and Simon Schaffer (1985: 24) observe that "physical hypotheses were provisional and revisable; assent to them was not obligatory, as it was to mathematical demonstrations; and physical science was, to varying degrees, removed from the realm of the demonstrative." In the experimentalist view, rejecting the demonstrative ideal of inquiry is a positive development.

"By the adoption of a probabilistic view of knowledge," Shapin and Schaffer (ibid.) continue, "one could attain to an appropriate certainty and aim to secure legitimate assent to knowledge claims." The problem with equating knowledge with impeccable certainty is dogmatism; the unwillingness to change accepted principles. Such an inappropriate objective was replaced by a fallibilistic account of knowledge. The probabilistic conception became embedded in experimentalism. Instead of axioms as the basis of physical propositions, Boyle argued that matters of fact are their foundation. To quote once again from Shapin and Schaffer (ibid.): "Boyle and the experimentalists offered the matter of fact as the foundation of proper knowledge. In the system of physical knowledge the fact was the item about which one could have the highest degree of probabilistic assurance: 'moral certainty.'"

Boyle's probabilistic-factualist account of knowledge is apparent in his 1660's work *New Experiments Physico-Mechanicall*, in which he reports the creation of a vacuum inside a glass globe. There are two salient epistemic criteria that the fact "there is a vacuum (or lower pressure) inside the confined space" requires: testimony of witnesses, and reproducibility. In the Introduction to the *New Experiments* ([1660]1999: 143), Boyle mentions the two criteria: "In divers cases I thought it necessary to deliver things circumstantially, that the Person I addressed them to, might without mistake, and with as little trouble as is possible, be able to repeat such unusual Experiments: and that after I consented to let my Observations be made publick." At the end of the quote, Boyle tells the reader that there need to be witnesses to the experimental procedure. The reliability of the testimony is dependent upon its multiplicity. This is a deliberate analogy between natural philosophy and criminal law (Shapin and Schaffer 1985: 56). In one of his often-quoted paragraphs, Boyle (1675: 182) notes that judging a person guilty or not guilty depends on the number of witnesses to the murder, and the agreement of different testimonies:

> For, though the testimony of a single witness shall not suffice to prove the accused party guilty of murder; yet the testimony of two witnesses, though but of equal credit … shall ordinarily suffice to prove a man guilty; because it is thought reasonable to suppose, that, though each testimony single be but probable, yet a concurrence of such probabilities, (which ought in reason to be attributed to the truth of what they jointly tend to prove) may well amount to a moral certainty, i.e., such a certainty, as may warrant the judge to proceed to the sentence of death against the indicted party.

The convergence of testimonial evidence increases the probability of a rightful conviction. The legal analogy—specifically, the reference to a substantial number of concurring witnesses increasing the probability of a rightful trial—renders the category of fact a social one. This is markedly different from Descartes' epistemology, which defines knowledge in terms of a lack of doubt, or indubitability (see Newman 2014). In his *Discourse on the Method*, Descartes (2000) describes a process in which a lone meditator can reach knowledge. For him, knowledge is something that is impossible to doubt. Impeccable certainty is attainable if an individual follows a proper method of reasoning. Boyle's experimentalist program is different, as it emphasizes the size and quality of the community testifying that the experimental procedure is carried out properly.

The second crucial criterion for probable knowledge is the number of consecutive testimonies. Hence repetition becomes relevant. In addition to the quality and number of witnesses, it is vital that the vacuum in the tube could be reproduced repeatedly. Thomas Hobbes immediately criticizes the experimentalist project in his 1661 treatise *Dialogus physicus de natura aeris* on this point. Experiments seem useless if they must be carried out over and over again. If someone could observe the relevant consequences in the experiment, for example, how different fluids behave under varying pressures, why can we not see it in a single experiment (Shapin and Schaffer 1985: 111)? In his *New Experiments*, Boyle describes in total forty-three consecutive experiments. These include experiments that first calibrate the machine and make sure it works properly. In the subsequent experiments, various objects, like flaming candles, or a bird (a lark), are placed inside the tube, and the consequences are observed.

Boyle's philosophy of experiment is relevant for understanding Hume's argument for the category of fact. Hume formulates the argument especially in his first *Enquiry*. Facts do not amount to demonstrations, but at most extremely high probabilities. They are socially corroborated, as they are dependent on testimonial evidence. Repetition, or frequent experience, is consequential for facts, because the number of experiences in favor of a factual proposition gives the basis for assessing its probability.

Before turning my attention to Hume's views, I will examine Newton's experimentalist methodology and rules of reasoning. This is important as Newton's approach stresses the importance of empirical evidence over hypotheses based on intuitive principles. In an analogous way, Hume's epistemology eschews *a priori* factual knowledge as matters of fact require experience.

Newton's Experimentalism and Criticism of Hypotheses

Boyle devised his experimentalism in the context of London's recently established Royal Society in the early 1660s and its precursor societies in Oxford in the late 1650s (MacIntosh and Anstey 2014: Section 6). For his part, Newton first stressed the mathematical treatment of natural philosophy. This is echoed by the title of his *Principia*: *The Mathematical Principles of Natural Philosophy*. Only when he responded to the criticism made by the mechanist and speculative philosophers in the General Scholium to the second edition of the *Principia* in 1712, did Newton call his enterprise experimentalist.

Newton's anti-hypothetical experimentalism is a polemical defense against Cartesian and Leibnizian natural philosophies. In this context Newton specifically defends one of the most important conclusions of his work, to wit, the law of universal gravitation (Shapiro 2004: 185). He advances an experimentalist argument against the criticism and the alternative mechanistic and speculative astrophysics. It is thus worth briefly explaining (cf. Chapter 1) what Newton's argument for the law of universal gravitation is and why it was so vehemently attacked. To put it concisely, Kepler's area law is the first premise of his argument, from which the conclusion, the law of universal gravitation, is deduced. By the method of demonstrative induction, and with the aid of background and structural assumptions, Newton argues that all the objects in the universe (in the scale of our solar system, at least) attract each other with a force that is directly proportional to the product of the masses and inversely proportional to the square of the distance between them. The Cartesian and Leibnizian philosophies are not necessarily in contradiction with the first and the second part of Newton's argument (see Belkind's 2012: 138–68 interpretation of Newton's argument).[5] In his article "Against Barbaric Physics," written around 1710–1716, Leibniz (1989a: 314) partly subscribes to Newton's argument. He even praises the inverse-square concept of the centripetal force, calling it a "beautiful discovery," and notes that "those who have shown that the astronomical laws can be explained by assuming the mutual gravitation of the planets have done something very worthwhile." His criticism focuses on the third, final part of the argument. Newton claims that the third law is a universal law of interaction; any body attracts every body with equal and opposite force. Thus Newton puts it as follows in Proposition 7 of Book 3 of the *Principia*:

> Further, since all the parts of any planet A are heavy [or gravitate] toward any planet B, and since the gravity of each part is to the gravity of the whole as the matter of that part to the matter of the whole, and since to every action (*by the third law of motion*) there is an equal reaction, it follows that planet B will gravitate in turn toward all the parts of planet A, and its gravity toward any one part will be to its gravity toward the whole of the planet as the matter of that part to the matter of the whole [my emphasis].

Leibniz argues that the universalization of Newton's third law, and hence the law of *universal* gravitation, lacks justification. By referring to Newtonians, he (1989a: 314) claims that "they even fabricate what they cannot prove through phenomena, for so far, except for the force by which sensible bodies move toward the center of the earth, they have not been able to bring forward any trace of the general attraction of matter in our region." Here we can see a partial approval of Newton's argument: "bodies move toward the center of the earth" because of the centripetal force. But Leibniz considers the "mutual gravitation of the planets" to be an unjustified explanation for astronomical motions. He notes that "we must be careful not to proceed from a few instances to everything" (ibid.). Newton was unable, Leibniz thinks, to provide empirical evidence for universal attraction that putatively exists between *every* particle in the universe.[6]

Although Leibniz notes that Newton might have lacked empirical evidence for his conclusion, the most pressing issue has to do with LUG's lack of reason. Universal gravitation violates Leibniz's principle of sufficient reason. In his correspondence with Clarke in 1713, Leibniz (1989b: 321) first explicates the principle of contradiction, and then adds that "in order to proceed from mathematics to natural philosophy, another principle is required [. . .] I mean the principle of sufficient reason, namely, that nothing happens without a reason why it should be so rather than otherwise."

The law of universal gravitation drastically violates Leibniz's principle. It is in numerous ways utterly strange, unintelligible, and unbelievable.[7] When considered from a paradigmatically mechanistic perspective, the law has the following astonishing aspects:

- The bodies change each other's motions through empty space.
- Gravity acts instantly and constantly, so cause and effect are contemporaneous.
- Gravity penetrates hard, putatively impenetrable matter, as it extends all the way to the center of bodies.
- Gravity's agent is left unspecified, so there is acceleration without an agent responsible for the change of motion.
- Gravity does not have a known underlying mechanism that explains how it is transmitted across space and time.
- Gravity is a force, so it is an interactive relation among at least two bodies (as Newton (1974a: 6) says, gravity law is "a branch of the third law of motion") and, consequently, an individual body does not generate a field that extends to infinity and which would be the causally efficacious agent for other body's motion.
- Gravity pertains to brute pieces of inert matter but causing an action at a distance seems like one piece of matter "knows" where the other piece of matter is.

In his 1693 correspondence with Richard Bentley, Newton (2004: 102) himself was very clear about the strangeness of the central result of his natural philosophy:

> It is inconceivable that inanimate brute matter should, without the mediation of something else, which is not material, operate upon and affect other matter without mutual contact [. . .] so that one body may act upon another at a distance through a vacuum without the mediation of anything else, by and through which their action and force may be conveyed from one to another, is to me so great an absurdity, that I believe no man who has in philosophical matters a competent faculty of thinking can ever fall into it.

Newton treats gravity as a real, causally efficacious force *and* thinks that instantaneous distant action is absurd. If one wants to maintain both premises—a realist ontology of forces and the admission that the way gravity causes motion is completely odd—the standards of natural philosophy must change.

In the General Scholium of the *Principia*, Newton tackles Leibniz's criticism by suggesting that although he was not able to provide any *reason* for the properties of gravity, he will not feign hypotheses. Hence Newton does not swallow the bait set by the principle of sufficient reason. If the law of gravity implies seemingly unintelligible properties, this point alone is not a valid counter-argument against it. A scientific principle must not necessarily comply with any intelligibility principle. Newton thinks that we should not give conditions to nature in what it ought to be, so that it would be intelligible to us.

As in the *Principia*, according to the methodology that Newton sets down in letters to Henry Oldenburg and Roger Cotes, "experimental philosophy proceeds only upon phenomena and deduces general propositions from them only by induction. And such is the proof of mutual attraction." Hypotheses should only be used to the extent that they have testable implications: "For hypotheses ought to be applied only in the explanation of the properties of things, and not made use of in determining them; except in so far as they may furnish experiments" (Newton 1974a, 1974b: 5–7). Newton therefore confines the domain of natural philosophy to statements that are deducible from the phenomena. Propositions that do not satisfy this criterion are mere hypotheses, and they are not acceptable in experimental science (Rutherford 2007: 12). He insists that principles in natural philosophy "are deduced from phenomena and made general by induction, which is the highest evidence that a proposition can have in this philosophy" (Newton 1974a: 6).

"The main business of natural philosophy," Newton (1979) proclaims in Query 28 of his *Opticks*, "is to argue from phenomena without feigning hypotheses." Newton is nevertheless willing to discuss provisional hypotheses (in his definition, propositions that are not deduced from the phenomena) that prompt future study of nature. This is essentially the purpose of the Queries: to formulate tentative, precursory questions so that other researchers in the future may make "a further search." In Query 21, Newton puzzles over the idea of there being an ether, a subtle fluid (something like, but less dense than air) that generates forces through the interaction of this medium's minute particles. Perhaps the space closer to the planets is loose (a small amount of corpuscles) and between the planets, around their common center of mass, the space is very dense (packed with corpuscles). Although Newton is sympathetic with a causal hypothesis like this, he explicitly denies having any knowledge about ether: "for I do not know what this aether is."

The ether hypothesis would provide a reason for gravity's operations, but there were no observations made, nor experiments carried out, that would have confirmed its existence. As Hylarie Kochiras (2011: 173–4) astutely remarks, acceptable hypotheses must be "amenable to empirical investigation." It is evident that, for example, Descartes' vortex theory is not subjected to such investigation. We do not see vortexes via telescopes, we are not able to collect that hypothetical substance and experiment with it, and so on.[8] Given the current situation of natural philosophy, in which Newton lacks empirical evidence for the cause of gravity, he admits at the end of the Queries that he does not want "to propose the principles of motion" from unobservable qualities "as they supposed to lie hid in bodies," but to "leave their cause to be found out."

Newton's theory of gravity does not make a reference to any mechanism that produces gravitational motions—it does not, in Newton's words, "unfold the mechanism of the world" (Opticks, Query 28)—but the mathematically characterized proportions of the law are close to the phenomena (*Principia*, Scholium, Book 1, Section 11, and Smith 2004: 11). According to Newtonian experimental philosophy, it is not requisite to reveal the supposedly intelligible essence of nature, such as the reason of gravity. Its objective is rather to mathematically describe gravity's effects and justify the resultant theory with experiments and observations.[9]

Newton is explicit in the contrast between the experimental and speculative/hypothetical philosophies of nature in his letter to Roger Cotes in 1713:

> Experimental philosophy reduces phenomena to general rules and looks upon the rules to be general when they hold generally in phenomena. It is not enough to object that a contrary phenomenon may happen but to make a legitimate objection, a contrary phenomenon must actually be produced. Hypothetical philosophy consists in imaginary explications of things and imaginary arguments for or against such explications, or against the arguments of experimental philosophers founded upon induction. The first sort of philosophy is followed by me, the latter too much by Descartes, Leibniz, and some others.
>
> Newton 2004: 120–1

In natural philosophy, it is not requisite to devise a hypothesis, like ether or vortex model, that forcefully makes the phenomena intelligible.[10] Rather, what matters is that general principles can be deduced from the phenomena inductively. "The central idea of the Newtonian inductive method," notes Graciela De Pierris (2006: 280), "is that exceptionless or nearly exceptionless universal laws are inductively derived from 'manifest qualities' or observed 'phenomena', and only further observed phenomena can lead us to revise these laws."

When Newton uses the term "induction" he is for the most part using it in the Aristotelian sense. In this respect, general principles are derived from particulars (Flew 1984: 171). Newton insists that he derived the law of universal gravitation from phenomena. Broadly speaking, this means that the law can be deduced (inferred in the mathematical, truth-preserving sense) from Kepler's empirical laws with the aid of several background assumptions (including structural assumption like momentum conservation, background assumption like Euclidean geometry, and approximating assumption like circular motion of satellites, Belkind 2012). Newton's argument involves drawing on a large data of astronomical observations as well as quantitative generalized laws of motion. David Marshall Miller (2009: 1052) notes that the *Principia* tries to establish that the qualities of bodies, and the three laws of motion and the law of gravity, "are features of every system of bodies in the universe."[11]

As this chapter aims to understand Hume's views in light of the preceding experimentalist tradition, Newton's Rule 3 in *Principia*'s Rules for the Study of Natural Philosophy can be seen as the most relevant background for Hume's experimentalism. As the rule is of central importance, it is appropriate to cite a significant amount of it:

> [T]he qualities of bodies can be known only through experiments; and therefore qualities that square with experiments universally are to be regarded as universal qualities [...] Certainly idle fancies ought not to be fabricated recklessly against the evidence of experiments, nor should we depart from the analogy of nature, since nature is always simple and ever consonant with itself. The extension of bodies is known to us only through our senses, and yet there are bodies beyond the range of these senses; but because extension is found in all sensible bodies, it is ascribed to all bodies universally. We know by experience that some bodies are hard [...] That all bodies are impenetrable we gather not by reason but by our senses.

In the rule above, Newton describes what we would today call Humean induction: extrapolation from the observed to the unobserved (Millican 2002: 112). In Janiak's (2007: 142) interpretation, Newton genuinely thinks that "a wide-range of previously disparate phenomena," such as the free fall of bodies and parabolic trajectories near the surface of the earth, the tides, the planetary and satellite orbits, and the orbits of distant comets from the earth, "have the *same* cause." For Newton, there is "no doubt that the nature of gravity toward the planets is the same as toward the earth" (*Principia*, Book 3, Proposition 6, Theorem 6). Gravity causes accelerations universally, effecting the wide-range of (both terrestrial and celestial) phenomena on the scale of our solar system. The universal force explains a good deal of its effects, but Newton does not explain gravity itself. He does not know the cause of this force: "Thus far I have explained the phenomena of the heavens and of our sea by the force of gravity, but I have not yet assigned a cause to gravity" (General Scholium, *Principia*).

There are many similarities between Newton's and Hume's positions: i) we know about the qualities/properties of bodies by experimentation and sensory evidence, and ii) we do not know about the qualities/properties of bodies based on our reason alone, so iii) we should not come up with non-experimental hypotheses about what the qualities/properties of bodies are. Ascription of qualities/properties iv) requires the assumption of uniformity of nature, that nature remains the same from past to future (as Newton says, nature is consonant to itself). In the next section, I will analyze these points of confluence between Newton and Hume (and points of divergence, too) along with Boyle's experimentalism.

Hume's Experimentalism

There is probably no better place to start with Hume's experimentalism than the Introduction to his *Treatise*. As the subtitle of the work indicates, Hume wants "to introduce the experimental method of reasoning into moral subjects." The subtitle does not mention the term "natural philosophy." However, Hume is certainly well aware that natural philosophy preceded the new moral philosophy: " 'Tis no astonishing reflection to consider, that the application of experimental philosophy to moral subjects should come after that to natural at the distance of above a whole century" (T Intro 7; SBN xvi– xvii). Given this admission, it seems clear that Hume wanted to emulate both the method and the reasoning of experimentalist natural philosophy. Moreover, in his view, the contemporaneous sciences were truly in a bad shape. "The systems of the most eminent philosophers" applied uncritically assumed principles which just aimed for a coherent world picture (T Intro 1; SBN xii). Hume considers this to be abstruse metaphysical thinking which should be replaced by a scientific investigation of our natures. The experimental approach is needed for the investigation of the principles of the human mind, as well as for improvement of the already existing sciences (Stanistreet 2002: 17–18).

In the Introduction to the *Treatise*, Hume does not mention Newton (or Boyle) by name.[12] In a Newtonian fashion, he does however refer to "a kind of Attraction" in the mental realm (T 1.1.4.6; SBN 12–13):

> Here is a kind of Attraction, which in the mental world will be found to have as extraordinary effects as in the natural, and to shew itself in as many and as various forms. Its effects are every where conspicuous; but as to its causes, they are mostly unknown, and must be resolv'd into *original* qualities of human nature, which I pretend not to explain.

Hume seems to invoke attraction to explain the orderliness and the structure of our thought. The copy principle explains how we come to have ideas that are the material of our thinking, but not how they are arranged. We could not think consistently, if ideas just occurred randomly, if our thoughts were "loose and unconnected." The imagination could separate and combine simple ideas randomly "were it not guided by some universal principles" (T 1.1.4.1; SBN 10–11; Morris and Brown 2014: section 4.3). As we can and do reason coherently, "chance alone wou'd join them." Instead, Hume finds three associative principles that are responsible for the regular order of our thoughts: 1) resemblance, 2) temporal and spatial contiguity, as well as 3) causation. In the first *Enquiry* (3.2; SBN 24), Hume thinks he discovered these universal principles in the same way as Newton discovered the law of universal gravitation. The principles of association are original; there is no further explanation as to why we associate ideas the way we do. The effects of these principles are apparent, although their causes "are mostly unknown [...] which I pretend not to explain" (T 1.1.4.1; SBN 10–11; Morris and Brown 2014, section 4.3). Here we can see an analogy with Newton's acknowledgement of the explanatory limitations of his universal dynamics.

In Stephen Buckle's (2004: 27–8) reading, in the Introduction Hume "takes up the prominent Newtonian theme that the philosopher must eschew 'hypotheses.'" Hume thinks that reasoning should stay within the bounds of experience, that is, within the results of "careful and exact experiments, and the observation of those particular effects" (T Intro 8; SBN xvii). Buckle continues: "The improvement of philosophy depends on being experimental in this sense, and does so because the hidden properties of things can never be known." For Hume, "the utmost extent" of human understanding is when we acknowledge "our ignorance, and perceive that we can give no reason for our most general and most refined principles, beside our experience of their reality" (T Intro 9; SBN xviii).[13] Very much like Newton, Hume wishes to avoid "that error, into which so many [speculative and hypothetical philosophers like Descartes and Leibniz] have fallen, of imposing their conjectures and hypotheses on the world for the most certain principles" (T Intro 9; SBN xix). Neither human nor natural science may explain any "ultimate principles." Even in their best theories, they both should refrain from going "beyond experience, or establish any principles which are not founded on that authority" (T Intro 10; SBN xviii). As Newton thinks that hypotheses, propositions not deduced from the phenomena, to use his language, "have no place in experimental philosophy," so Hume also asserts that "any hypothesis, that pretends to discover the ultimate original qualities of human nature, ought at first to be rejected as presumptuous and chimerical" (T Intro 8; SBN xvii).

In my view, the Introduction to the *Treatise* is particularly but not exclusively Newtonian. Steffen Ducheyne (2009) and Eric Schliesser (2009) challenge the Newtonian reading. They point out that the label "experimental" was common in many textbooks of natural philosophy in Hume's time. The entry of "Experimental" in Chambers' dictionary (1728: 368) mentions, besides Newton, figures like Bacon and Boyle, and also the institutions Accademia del Cimento, the Royal Society, and the Royal Academy of Paris. Schliesser (2009: 173) makes the specific point that Hume's endorsement of experimentalism is closer to Boyle's scientific methodology than Newton's.

I think there are two points in common between Newton's and Hume's experimentalisms. First, Hume is especially critical about hypotheses (specifically in T Intro 9; SBN xix). In his other writings (EHU 7.25; SBN 73, fn. 16, *History*: VI, 542), Hume clearly recognizes that it was *Newton* who criticized hypotheses on the basis of his experimentalist methodology. Critique of hypotheses in the context of experimentalist methodology is a particularly Newtonian thought (Anstey and Vanzo 2012: 518). Hume writes that:

> It was never the meaning of Sir Isaac Newton to rob second causes of all force or energy; though some of his followers have endeavoured to establish that theory upon his authority. On the contrary, that great philosopher had recourse to an etherial active fluid to explain his universal attraction; though *he was so cautious and modest as to allow, that it was a mere hypothesis, not to be insisted on, without more experiments.*
>
> EHU 7.25; SBN 73, fn. 16; my emphasis

In Newton this island may boast of having produced the greatest and rarest genius that ever arose for the ornament and instruction of the species. *Cautious in admitting no principles but such as were founded on experiment; but resolute to adopt every such principle, however new or unusual:* From modesty, ignorant of his superiority above the rest of mankind; and thence, less careful to accommodate his reasonings to common apprehensions.

> History VI: 542; my emphasis

Hume's admiration (and nationalist, propagandist presentation) of Newton's experimentalism does not of course exclude his affinity for Boyle. In his *History* (VI: 541), Hume's rhetoric is appreciative of Boyle: "Boyle improved the pneumatic engine invented by Otto Guericke, and was thereby enabled to make several new and curious experiments on the air as well as on other bodies: His chemistry is much admired by those who are acquainted with that art." Hume clearly reveres Boyle's experimental pneumatics. Nevertheless, Hume goes on to reject Boyle's corpuscularian theory of matter (or at least he remains agnostic on its existence). In the same paragraph, Hume suggests that Boyle went too far in providing an explanation for the operation of air by positing an imperceptible microstructure for matter: "Boyle was a great partizan of the mechanical philosophy; a theory, which, by discovering some of the secrets of nature, and allowing us to imagine the rest, is so agreeable to the natural vanity and curiosity of men."

Second, Newton's inductive method of reasoning as he articulates it in Rule 3 is very congenial to Hume's experimentalism. Newton argues that the qualities of bodies can be known only by experimentation and inductive inference, not by *a priori* assumptions of primary qualities of matter (see De Pierris 2006). As I have already cited Rule 3 at length in the previous section, there is no need to reproduce it here. It suffices to note that Newton's contentions, such as "the qualities of bodies can be known only through experiments," and "idle fancies ought not to be fabricated recklessly against the evidence of experiments," closely resemble Hume's related views (ibid.: 282–3).

Despite these similarities, there is one crucial difference between Hume and Newton. In Rule 3, Newton mentions the force of inertia, that is, one definition of mass. Although in Rule 3 Newton denies reason's capacity to tell us about the properties of objects, he still maintains that rational inference is a source of knowledge about properties (Miller 2009: 1054). This is apparent in Newton's own words in the very same third rule: "That all bodies are movable and persevere in motion or in rest by means of certain forces (which we call forces of inertia) we infer from finding these properties in the bodies that we have seen." This is an important difference between Newton's and Hume's epistemologies and ontologies, because Newton takes imperceptible items, like mass, into his ontology, whereas Hume is adamant that inquiry should stay strictly within the bounds of impression-based ideas and experience (see especially T 1.2.5.26; SBN 64, fn. 12).

Moreover, Hume notes that experiments in human science do not involve predictions and artificially created circumstances. Experiments in moral philosophy are not done "purposely, with premeditation" (T Intro 10; SBN xviii–xix, and Demeter 2012: 582). By manipulating salient variables like friction and air resistance, physical experiments create artificial scenarios that would not exist without human intervention. It is obvious that one cannot intervene in historical events in the same way as in physical events, like pressure differences of fluids. The moral philosopher observes the uninfluenced behavior of other people in the ordinary courses of their lives. This is the way data collection and experimentation are done in the science of humanity. Hume's imperative is that: "We must therefore glean up our experiments in this science from a cautious observation of human life, and take them as they appear in the common course of the world, by men's behaviour in company, in affairs, and in their pleasures" (T Intro 10; SBN xviii–xix). By carefully collecting and comparing such experiments, the new moral science will be made possible. It will not be less certain than natural sciences, and it will be more useful than any other form of human inquiry (Stanistreet 2002: 20).

On a further note, Hume criticizes Newton, or a Newtonian argument, on a wrongful use of analogical reasoning. The Newtonian natural philosophy in Scotland was tightly connected with natural theology in Hume's time. He knew about Newton's design argument, plausibly from Maclaurin's *An Account of Sir Isaac Newton's Philosophical Discoveries* (Hurlbutt 1965: 42). The orderliness and complexity of the natural world gives a basis, Hume's Newton thinks, to infer the existence of God (Stanistreet 2002: 36).[14] Hume presents a critique of the argument in the discussion of Philo and Cleanthes in the *Dialogues*. Cleanthes purports to establish *a posteriori* the analogy of the natural world and its Creator. He argues that the world is comparable to a human-made machine, which resembles and exceeds "the productions of human contrivance; of human design, thought, wisdom, and intelligence" (DNR 2.5; SBN KS 143). Because "the effects resemble each other, we are led to infer, by all the rules of analogy, that the causes also resemble." Machines have makers, and as nature is machine-like, it also has a maker. The corollary is "that the Author of Nature is somewhat similar to the mind of man; though possessed of much larger faculties, proportioned to the grandeur of the work, which he has executed. By this argument *a posteriori*, and by this argument alone, do we prove at once the existence of a Deity, and his similarity to human mind and intelligence" (ibid.)

Hume, in the line of Philo, criticizes the argument because the analogy here is of a weak nature (DNR 2.7; SBN 144). He also makes a similar argument in the first *Enquiry* (9.1; SBN 104–5), when addressing the common points between human and animal reasoning. Hume acknowledges the importance of analogical reasoning: "All our reasonings concerning matter of fact are founded on a species of Analogy, which leads us to expect from any cause the same events, which we have observed to result from similar causes." He adds that in perfectly similar cases, analogical reasoning is trustworthy: "Where the causes are entirely similar, the analogy is perfect, and the inference, drawn from it, is regarded as certain and conclusive." By means of analogy we can conclude several similarities with certainty, like hardness of macroscopic bodies and blood circulation in animals.[15] The evidence in favor of the design argument, however, is considerably weaker. In the analogy, the very things that are compared are connected more loosely. Hume, in the voice of Philo, debunks Cleanthes' analogy (DNR 2.8; SBN 144):

> If we see a house, Cleanthes, we conclude, with the greatest certainty, that it had an architect or builder: because this is precisely that species of effect, which we have experienced to proceed from that species of cause. But surely you will not affirm, that the universe bears such a resemblance to a house, that we can with the same certainty infer a similar cause, or that the analogy is here entire and perfect. The dissimilitude is so striking, that the utmost you can here pretend to is a guess, a conjecture, a presumption concerning a similar cause; and how that pretension will be received in the world, I leave you to consider.

Stanistreet (2002: 37) notes that Hume's criticism of the design argument in the first *Enquiry* and *Dialogues* relates to his doctrine of relations in the *Treatise*. When a resemblance relation between two species of objects becomes weaker, then also the probability of the analogy that the statement tries to establish diminishes (T 1.3.13.8; SBN 147): "In proportion as the resemblance decays, the probability diminishes; but still has some force as long as there remain any traces of the resemblance." Finally, Hume disagrees with Newton's argument on experimental grounds. Although his empiricism is indebted to the experimentalist tradition, the design argument leans on unobservable causes. As Philo contends (DNR 2.13; SBN 145–6): "Experience alone can point out [...] the true cause of any phenomenon."

This section has established that Hume, at least on a rhetorical level, subscribes to the experimentalist tradition of natural philosophy, although there are some nuanced differences. Next, I will focus on Hume's argument for the category of matters of fact, and how it is enmeshed in experimentalism. Hume argues that reasoning concerning facts is probabilistic, and that we identify causal propositions with experience. Causation is drastically important in this context, but as I will address it in the subsequent fourth section (in tandem with laws of nature and metaphysics of forces) I shall restrict my treatment to experience and highlight the social and probabilistic nature of matters of facts. I wish to show that this places Hume firmly in the tradition of British experimental philosophy.

Matters of Fact: Experience, Testimony, and Probability

Hume sketches his argument for the category of fact in the fourth section of his first *Enquiry*. The section introduces Hume's fork, a distinction rooted in the doctrine of relations in *Treatise* 1.3.1. The two propositions, relations of ideas and matters of fact, are categorically different. Hume divides the two in terms of conceivability of their negations.[16] Although the divide is needed to understand both categories of knowledge, I will here focus solely on facts as I wish to explore Hume's argument by considering experimental philosophy.

In accordance with Boyle's experimentalist position, Hume rejects the demonstrative certainty of matters of fact. In the first paragraph of the fourth section of the first *Enquiry* (4.1; SBN 25), Hume argues that Euclidian geometry can attain perfectly certain knowledge: "the truths, demonstrated by Euclid [. . .] would for ever retain their certainty and evidence." Right after this sentence, Hume continues to argue that "matters of fact, which are the second objects of human reason, are not ascertained in the same manner; nor is our evidence of their truth, however great, of a like nature with the foregoing" (EHU 4.2; SBN 25–6).

Mere perception, say, a perception of an individual idea, or a perception of ideas standing in some relation, does not convey to us information about any factual matter. Hume therefore asks: "what is the nature of that evidence, which assures us of any real existence and matter of fact" that is not immediately present to the senses, or of which we do not have a memory trace of? In short, the answer is causation. As causation is the subject of Chapter 4, we may here address Hume's answer to the subsequent problem: How do we acquire knowledge concerning causal relations? Answering this question is also solving the underlying problem about "the nature of that evidence, which assures us of matters of fact" (EHU 4.3–5; SBN 26–7). Hume is very clear in his answer: "I shall venture to affirm, as a general proposition, which admits of no exception, that the knowledge of this relation is not, in any instance, attained by reasonings *à priori*; but arises entirely from experience, when we find, that any particular objects are constantly conjoined with each other" (EHU 4.6; SBN 27).

To further amplify this point, Hume makes use of a thought experiment involving an ideal but *a priori* epistemic agent. By referring to previous medieval tradition,[17] he invokes Adam, a person with perfect sensory system and reasoning faculties, who was just created and immediately brought into the world. Adam's cognitive capacities are otherwise superior, but he lacks experience. He does not have any information about the uniformity of nature (roughly, that the future conforms to the past). Now imagine that Adam faces, for the very first time, the following objects in front of him: water, flame, gunpowder, and lodestone. Does he know, based on his perception of the objects, and on his reasoning, that water may cause drowning to non-aquatic beings, that flame is the cause of heat, that setting up fire causes powder to explode, and that lodestone attracts metallic pieces? Hume's answer is an explicit and categorical no: "Let an object be presented to a man of ever so strong natural reason and abilities; if that object be entirely new to him, he will not be able, by the most accurate examination of its sensible qualities, to discover any of its causes or effects" (EHU 4.6; SBN 27). Although Adam's senses and reason are similar to ours—more precisely, they are as good as they get—he lacks the experience we have. As he lacks everyday experience, he cannot have any natural scientific knowledge either. He could know mathematical truths, provided that he gets the relevant arithmetic, algebraic, and geometric ideas like numbers, variables, and figures before doing any computations or proofs. But he could not have any clue of what kinds of laws operate in nature. Take, for example, the Galilean–Newtonian law of free fall of objects, or the Cartesian–Newtonian law of conservation of momentum. Hume (EHU 4.9; SBN 29) uses these examples "to convince us, that all the laws of nature, and all the operations of bodies without exception, are known only by experience."

Adam has a piece of metal in his hand. He releases it. Does he know that it will fall downwards? He has never seen a motion like that before. What reasons does he have for such arbitrary conjecture? Why is it more reasonable to say that the metal falls instead of saying that it goes up, stays put, or disappears and instantly appears somewhere else? Or, what would Adam think if he were to witness a game of pool next to a pool table? He sees the cue ball moving toward the object ball. Does he know the resultant motion of the balls? What reasons does he have for surmising that the cue ball comes to a halt and the object ball continues with roughly the same motion? Why would he not guess that both balls stop completely, or jump off the table, or initiate a backward motion? Adam has only his immediate sensory evidence and reasoning faculty, but no experience and no assumption about the uniformity and regularity of nature. In the Abstract (11–14; SBN 650–2) to his *Treatise*, Hume argues that Adam does not know beforehand what happens after the collision of objects:

Were a man, such as *Adam*, created in the full vigour of understanding, without experience, he would never be able to infer motion in the second ball from the motion and impulse of the first. It is not any thing that reason sees in the cause, which makes us *infer* the effect. [...] It would have been necessary, therefore, for *Adam* (if he was not inspired) to have had *experience* of the effect, which followed upon the impulse of these two balls. He must have seen, in several instances, that when the one ball struck upon the other, the second always acquired motion. If he had seen a sufficient number of instances of this kind, whenever he saw the one ball moving toward the other, he would always conclude without hesitation, that the second would acquire motion. His understanding would anticipate his sight, and form a conclusion suitable to his past experience. [...] It follows, then, that all reasonings concerning cause and effect, are founded on experience, and that all reasonings from experience are founded on the supposition, that the course of nature will continue uniformly the same. [...] 'Tis evident, that *Adam* with all his science, would never have been able to *demonstrate*, that the course of nature must continue uniformly the same, and that the future must be conformable to the past.

Mere *a priori* reasoning is completely impotent when predicting the motions of bodies. There is an indefinite number of logical possibilities as to which direction objects move after the application of forces. Even more broadly, it is not *a priori* clear how an object even exists after the application of forces: does it compress, remain as it is, or disappear? Hume thinks Adam has no reason whatsoever to decide in which direction a metallic piece will fall immediately after being released, or how colliding billiard balls spread out on the table: "All these suppositions are consistent and conceivable. Why then should we give the preference to one, which is no more consistent or conceivable than the rest? All our reasonings *à priori* will never be able to shew us any foundation for this preference" (EHU 4.10; SBN 30).

Just by inspecting different elements, such as rocks, magnets, or water, Adam does not gain access to their respective qualities. The only way for us, or even an ideal epistemic agent, to know about qualities of objects is by experimenting with them. In Hume's terminology, we need to have frequent experience of the constant conjunctions between species of objects or events. Thus Hume defines the term "experience":

Tis therefore by experience only, that we can infer the existence of one object from that of another. The nature of experience is this. We remember to have had frequent instances of the existence of one species of objects; and also remember, that the individuals of another species of objects have always attended them, and have existed in a regular order of contiguity and succession with regard to them.

T 1.3.6.2; SBN 87

Hume's reference to frequent experience[18] also has a social dimension. This is evident in section ten of the first *Enquiry* (10.3; SBN 110, and 10.5; SBN 111–12): "Though experience be our only guide in reasoning concerning matters of fact; it must be acknowledged, that this guide is not altogether infallible [...] there is no species of reasoning more common, more useful, and even necessary to human life, than that which is derived from the testimony of men, and the reports of eye-witnesses and spectators."

Propositions concerning facts are social. Laws of nature, for example, are witnessed by the whole of humankind. In Hume's view, the proposition "water is potentially drowning to non-aquatic beings" is a law of nature. It is evident that such a law could not be known based on individual experiences. I cannot go on and test whether this proposition is true or false. Inductive support for propositions is dependent on the multiplicity of testimonies. In assessing the probability of factual propositions, the number and quality of testimonies are relevant. Even the most reliable eyewitness may err in testimony, whether the eyewitness is mistaken by accident or lying deliberately. However, if the testimony is multiplied with a number close to the whole human population, it seems that such individual mistakes will gradually be eliminated. There will be convergent evidence in favor of laws of nature and contrary evidence to testimonies about violations of laws of nature. For Hume, testimonial epistemology relates to his critique of reported religious miracles. In this regard, he diverges from Boyle. For the latter, testimony of miracles served as proof, not denial of them. In Boyle's case, testimonies were important for the fulfilling of biblical prophesies. This is "radically contrary to basic Hume," notes Eugenie Sapadin (1997: 320).

Given the social-testimonial nature of the evidence for factual propositions, we may assess conditional probabilities of propositions.[19] Given the indemonstrability of the uniformity principle, even laws of nature are not demonstrable, so their probability is less than 1. To use Hume's example, it is possible that a piece of metal could be left suspended in the air after releasing it. Nevertheless, the probability of a violation to a law like this can be approximated to be around 10^{-1000}.[20] Say that a reliable eyewitness is right 9,999 times out of 10,000; her probability of getting the testimony wrong would be 10^{-4}. Comparing the prior probability for the law happening as usual, and the probability of the witness getting the testimony right, even the very best eyewitness who testifies that a metallic piece released from the hand did not fall, yields a probability $\approx 10^{-1000}$ for the miraculous event to occur. The initial probability for a law to not take place as usual is extremely small.

Here we may again see Hume's debt to the experimentalist tradition of the early modern natural philosophy. As Boyle had premised before Hume, certainty is a matter of degree. In the section "Of Probability" of his first *Enquiry*, Hume distinguishes three degrees of certainty: demonstration, proof, and probability (EHU 6 fn. 10; SBN 56). According to this tripartite definition of knowledge, facts are either probable or provable, whereas mathematical propositions are demonstrable. Consider the three following propositions: "If I take a painkiller, that will relieve my headache"; "If I release a metal piece from my hand, it will fall"; "The sum of the angles of a triangle equals two right angles." In the first two sample propositions, the probability is dependent on a number of previous observations. In the case of taking a painkiller, it sometimes but not always has relieved my pain. But every time I have released a metallic object it has fallen from my hand. There is no exception for this fact in our past experience. Hume (EHU 10.3–4; SBN 110–11) describes the difference between proofs and probabilities:

> Some events are found, in all countries and all ages, to have been constantly conjoined together [proof]: Others are found to have been more variable, and sometimes to disappoint our expectations [probabilities]; so that, in our reasonings concerning matter of fact, there are all imaginable degrees of assurance, from the highest certainty to the lowest species of moral evidence. A wise man, therefore, proportions his belief to the evidence. In such conclusions as are founded on an infallible experience, he expects the event with the last degree of assurance, and regards his past experience as a full *proof* of the future existence of that event. In other cases, he proceeds with more caution: He weighs the opposite experiments: He considers which side is supported by the greater number of experiments: To that side he inclines, with doubt and hesitation; and when at last he fixes his judgment, the evidence exceeds not what we properly call *probability* [my additions in square brackets].

We do not doubt that laws remain the same, but as the uniform course of nature might change, so laws of nature might change in the future, too. That is why provable propositions concerning laws of nature are non-necessary: they might be otherwise. This is not the case with the truths of pure mathematics. Propositions like Euclid's sum-angle theorem can be demonstrated. They cannot be otherwise in the sense that their negations are inconceivable, and they do not vary across space and time. But this will be the topic of Chapter 5; next I shall scrutinize Hume on laws, causation, and the ontology of forces.

Laws of Nature, Causation, and the Ontology of Forces

This chapter, as its title suggests, deals with three interrelated concepts. Laws of nature, causation, and the problem of the existence of forces are tightly connected themes. To analyze Hume on these three themes, the rest of this chapter is composed as follows. In the next section, I trace the history of the concept of a law of nature and explicate the two main positions on the metaphysics of laws in the early modern period. Then I show that for Hume laws are causal. Thereafter I present my reading of Hume on causation. In accordance with the traditional regularity theory interpretation, I argue that causation is a discovered constant conjunction. What is discovered is the regular, unexceptional relation between species of objects or events. Finally, I investigate the notion of mechanism, the putatively causally efficacious part of bodies, and its relation to Hume's concept of causation. I argue that Hume tacitly assumes a mechanism in his reference to the laws of physics. Unlike Boyle and Locke, he does not think that we have discovered any necessitating causal power. The concept of force is still meaningful for Hume, because it is an instrument or calculating device which enables one to predict the various outcomes of motions of bodies.

The Notion of a Law of Nature

The world is not a chaos. We see bodies moving according to regular patterns. Why is this so? In the first half of the seventeenth century, the concept "law of nature" was introduced to explain the stability of the universe. The then dominant "Cartesian theory of the world," Catherine Wilson (2008: 25) clarifies, "gives us no reason to believe that our present animals, vegetables, and minerals—or even creatures with human bodies like ours—or for that matter, oceans and mountains like ours, have been here for any length of time and will survive into future epochs rather than being replaced by other creatures and features that are mechanically possible."[1]

Before Descartes, the notion of a natural law was used primarily in the context of the divine command theory. According to this theory, praiseworthy actions are those that follow the will of God, and blameworthy actions are those that contradict God's will. Descartes translated this theological concept to the realm of bodies. Thus laws of nature, in addition to bodies, are the subject matter of physics (Ott 2009: 1). In his *Principles* (II 37), Descartes argues that his laws of motion are derivable from God's immutability. Laws, including the three set down in the *Principles*, are the secondary "causes of the diverse movements which we notice in individual bodies." The principle of inertia, for example, is a consequence of God's unchangeability: bodies stay in their state of motion unless there is an external cause that changes their state (Ott 2009: 5). God is also responsible for the conservation of momentum: "He conserves motion; not always contained in the same parts of matter, but transferred from some parts to others depending on the ways in which they come in contact. Thus, this continuous changing in created things is an argument for the immutability of God" (Pr II 42).

In his 1685 work, *Free Inquiry into the Vulgarly Received Notion of Nature*, Boyle criticized Descartes' law metaphor. It seems strange to say that bodies could obey or disobey any laws or rules. In Descartes' metaphysics, humans are mental beings, but the rest of nature is machine-like automata. Bodies are extended beings, inert matter. The substance of body is exhausted by its attribute of extension and its contingent mode. Bodies cannot be governed in any way: they are not "allowed" or "forbidden" to do anything. They just move due to causes external to them by collision impact. Only beings capable of moral reflection, to wit, humans, may be commanded by God. Although Boyle rejects Descartes' position on laws of nature, he still seems to accept some concept of a law of nature. In the *Free Inquiry*, we see that Boyle (1685: 40–1) is hesitant, but does not want to eliminate the usage of the word "law" in natural philosophy:

> There is oftentimes some Resemblance between the orderly and regular Motions of inanimate Bodies and the Actions of Agents, that, in what they do, act conformably to Laws. And even I sometimes scruple not to speak of the Laws of Motion and Rest that God has established among things Corporeal, and now and then to call them, as men are wont to do, the *Laws of Nature*.

According to Walter Ott (2009: Section 1.1), we may roughly distinguish two prominent positions on the metaphysics of laws of nature among early moderns: the top-down and the bottom-up analyses. Descartes' position is the top-down analysis. Laws are independent of bodies, and they make bodies do what they do.

Laws are not fixed by the bodies they govern. They are not dependent on created beings but solely on God. This analysis proceeds from the general (laws) to the particulars (bodies). The alternative bottom-up analysis is shared by Boyle, Locke, and Hume: Laws describe the interrelationships between salient quantities like mass, distance, and velocity of bodies. However, there are no things called "laws" that somehow float free of bodies and prescribe their behavior. This analysis proceeds from the particulars (the bodies' properties and quantities) to the general (relations among the properties and quantities). Laws of nature are nothing more than convenient ways of stating the relations among the relevant properties and quantities. "Nature takes the course it does simply in virtue of the kind of things that make it up," Ott (2009: 6) explains.

The bottom-up analysis maintains that laws are not extrinsic to physical objects or events. To quote Jonathan Schaffer (2008: 82), "the laws of nature are nothing over and above the pattern of events, just like a movie is nothing over and above the sequence of frames." The regular relationship, or, more precisely, a generalization over the regularities, is what we call the laws of nature. For example, igniting a pile of gunpowder regularly results in an explosion. Based on this regularity, it may be generalized that under such and such circumstances, all piles of powder explode after certain reactions.

Given the former historical exposition, my assumption is that Hume's position is congenial to the bottom-up analysis. Laws do not govern the behavior of bodies but propositions concerning laws are informative of how bodies move. This leads to a potential difficulty. In Hume's view, *laws do not cause* anything, but he still maintains that *laws are causal*. The next section provides evidence that Hume does indeed hold laws to be causal. I wish to show that there is nothing paradoxical about this.

Laws are Causal

Laws of nature are matters of fact. Propositions concerning matters of fact are causal (T 1.3.1.1–3; SBN 69–70). Hume makes a universal claim that "all reasonings concerning matter of fact seem to be founded on the relation of *Cause and Effect*" (EHU 4.4; SBN 26). Laws of physics are generalized causal principles, as they are informative of what sorts of effects (body motions) are to be expected in what sorts of antecedent circumstances. Reasoning concerning laws is accordingly founded on the same factors that establish other causal principles: association based on past experience of conjunctions between types of events. In a word, laws of nature are causal and informative of how bodies move.

In Hume's analysis of physical laws, causal language is rampant. It should be noted that he does not use dynamic concepts, like force, in an exact way. Occasionally, he substitutes it with moment (momentum), power, and energy. When addressing dynamic concepts like gravity, force, power, energy, and momentum, Hume uses active words; that is, verbs like cause, produce, impel, fix, determine, and make. The causal analysis of laws is especially transparent regarding the law and the force of gravity. The law belongs to "the most establish'd and uniform conjunctions of causes and effects" (T 1.3.8.14; SBN 104); the force is a cause "which determine[s] it [an object] to fall" (T 1.3.11.10; SBN 128); it is among "the ultimate causes and principles which we shall ever discover in nature" (EHU 4.12; SBN 30). "The production of motion by ... gravity is an universal law" (EHU 6.4; SBN 57), as the moon is kept "in its orbit by the same force of gravity, that makes bodies fall near the surface of the earth" (EMP 6.6; SBN 236).

In the first *Enquiry* (1.15; SBN 14–15), Hume even seems to lean toward a governing conception of laws. When he comments on (probably Newton's) contributions to astronomy, he notes that the motions of planets are governed by laws and forces: "Astronomers had long contented themselves with proving, from the phænomena, the true motions, order, and magnitude of the heavenly bodies: Till a philosopher, at last, arose, who seems, from the happiest reasoning, to have also determined the laws and forces, by which the revolutions of the planets are governed and directed."

The above quote is somewhat unclear on whether forces *or* laws govern bodies. From the viewpoint of the metaphysics of laws of nature, the two are not the same. NonHumeanism maintains that laws govern and restrict bodies' behavior. Humeanism denies that laws cause anything, but a Humean may still say that a force causes change of motion. The Humean maintains that bodies and their regular motions are all there is to causation in laws; beyond the bodies there are no laws that cause the bodies to do anything. It is not surprising that Hume understood laws of nature in causal terms. Both Cartesian and Newtonian natural philosophies are essentially causal in their make-up. Hume is not a critic of causality in a sense that he would like to eliminate causation like Berkeley in his *De Motu*, or like some subsequent philosophies of physics.[2]

As it has been established that laws are causal, the next task is to figure out what is causation for Hume. To that end, I shall first lay down my interpretation of Hume on causation. Then I will scrutinize the notion of causal power, which roughly equates to the mechanism of bodies.

Causation as Discovered Constant Conjunction

. . . the constant conjunction of objects determines their causation . . .

T 1.3.15.1; SBN 173

. . .causes and effects are discoverable. . .

EHU 4.7; SBN 28

Regarding causation, it is possible to separate three different interpretations: the traditional, reductionist interpretation (or the old Hume interpretation), projectivism, and skeptical realism (the new Hume interpretation). Helen Beebee (2006: 108) encapsulates the traditional interpretation by introducing a positive and a negative claim:

> The positive claim is that Hume holds that causation in the objects is a matter of temporal priority, contiguity and constant conjunction: our causal talk and thought cannot succeed in describing or referring to any more in the world than these features. The negative claim is that it is illegitimate or incoherent to apply the idea of necessary connection to external events.

The projectivist and the skeptical realist interpretations challenge this traditional interpretation. Projectivism denies the negative claim, while skeptical realism denies the positive. In the formulation of Don Garrett (2015b: 173–4), the three separate interpretations can be summarized as below:

- *Reductivism.* Causation is a constant conjunction of species of objects/ events; it is nothing over and above regular spatiotemporal relations.
- *Realism.* There are mind-independent causal powers which govern and necessitate the behavior of objects/events.
- *Projectivism.* Causal thinking denotes our habits of reasoning; we project our inferences from cause to effect onto the world as if it contained causal efficaciousness and necessity.

In her defense of a quasi-realist interpretation, Angela Coventry (2006: 3) develops something like a combination of the three readings. Her premise is this: "Currently, there are two opposing schools of interpretation: causal realist interpretations and causal anti-realist interpretations." The former view attributes truth-values to causal statements due to necessitating causal powers that bring about their effects. According to this view, causation is something fundamental in the universe. The latter view balks any realist-sounding claims concerning causal relations, because there is no evidence of causal powers that would ground our causal judgments. Causal statements express regularities in nature, no more. Coventry defends a middle position between the two by favoring projectivism. In making causal claims, we articulate our stances, attitudes, prescriptions, or desires. For Coventry, the regularity of nature cannot be explained by referring to an underpinning mind-independent causal connection. We can, however, explain and justify realistically motivated causal talk because we are thoroughly acquainted with all sorts of regularities (Coventry 2006: 65). Beebee's (2006: 10) reading is similar in spirit, as it emphasizes the projection of necessary connection onto the world.

If the original premise is a dichotomy between realism and anti-realism about causation, then our interpretations of Hume will lead us to consider one of the three received views: reductivism, realism, or, a sort of combination of them, projectivism. There is textual evidence for all of these readings, and it is unclear whether Hume even has a single theory of causation. And some positions might indeed combine aspects of different interpretations. In what follows, I argue for a version of the traditional regularity reading of Hume on causation. In this version, causal relations are discovered in our natural environment. What is discovered, however, is nothing more than constant conjunction.

A regularity theory of causation is consistent with some aspects of causal realism. There is good textual evidence to indicate that Hume is a realist about regularities. Before making our argument and marshalling the evidence, we should be careful with the definitions of the terms. In the formulation of Michael Esfeld (2011: 157), causal realism is the following position: "causation is a fundamental feature of the world, consisting in the fact that the properties that there are in the world, including notably the fundamental physical ones, are dispositions or powers to produce certain effects."

A similar metaphysically realist position is formulated by Rani Lill Anjum and Stephen Mumford. In their view, powers, or equivalently dispositions, are the causally efficacious factors. They are indispensable for causation. Anjum and Mumford (2011: 7) explain how their causal dispositionalism differs from that of Hume:

> It does not offer to replace causation with something else. Hume had such a project, for instance, arguing that when event *a* causes event *b* it means simply that *a* occurs before *b*, is contiguous with *b*, and that every *a*-like thing is followed by a *b*-like thing. Causation in his theory is thus reduced to the non-causal notions of temporal priority, spatial contiguity and constant conjunction.

If causal talk is reduced to mere spatiotemporal regularities, causal statements lack truth-values (Tooley 1987: 246). In other words, a view that reduces causation to something else cannot be realist. Stathis Psillos points out that such characterization is problematic. One can "accept that it is regularities *all the way down*, and yet also accept that these regularities are *real, objective and mind-independent*," he states (Psillos 2002: 23). The fact that causation reduces to regularity does not make it disappear. One can say consistently that regularities are real, even though we should remain agnostic about the causally efficacious factors of bodies and natural events.

Below I argue that although Hume does not find any evidence of causal powers/dispositions, he still thinks that causation is *discovered* constant conjunction. The regular relations between species of objects/events are out there in our physical environment, waiting to be found out. Hume does not postulate causal attributes. We do not project a magnet's property to attract, or gunpowder's explosiveness. Rather, we identify such causal relations with frequent and collective experience. But we will not find a necessary connection among causal relations. The key to this view is that causation is regularity all the way down.

According to Hume's methodology, one should first inquire into the idea of causation. In the *Treatise*, he follows a precise methodology. He starts with a search for the idea of causation. Ideas are impression-based thoughts, the material of our thinking. Without ideas there is no reasoning. To explain causal reasoning, we must therefore be acquainted with the origin of our idea of causation. When we examine the sensory qualities of objects, we cannot find anything like cause and effect in them (T 1.3.2.5; SBN 75). Consider Hume's favorite example, billiard balls on the pool table. Looking at them, I see two colored, discrete objects. My perception of these objects, or their states of motion, does not tell me anything about causality. It must be that causation is a *relation* pertaining to these objects. Hume discerns two aspects of this relation: contiguity and temporal priority. When two bodies are causally related, they touch each other, and the cause comes before the effect in time (T 1.3.2.6–7; SBN 75–6). "Having thus," Hume writes,

> discover'd or suppos'd the two relations of *contiguity* and *succession* to be essential to causes and effects, I find I am stopt short, and can proceed no farther in considering any single instance of cause and effect. Motion in one body is regarded upon impulse as the cause of motion in another. When we consider these objects with the utmost attention, we find only that the one body

approaches the other; and that the motion of it precedes that of the other, but without any sensible interval. 'Tis in vain to rack ourselves with *farther* thought and reflection upon this subject. We can go no *farther* in considering this particular instance.

T 1.3.2.9, SBN 76–7

In the Abstract to the *Treatise*, Hume first recounts the requirements of contiguity and temporal priority. Then he adds that there

> is a *third* circumstance, *viz.* that of a *constant conjunction* betwixt the cause and effect. Every object like the cause, produces always some object like the effect. Beyond these three circumstances of contiguity, priority, and constant conjunction, I can discover nothing in this cause. The first ball is in motion; touches the second; immediately the second is in motion: and when I try the experiment with the same or like balls, in the same or like circumstances, I find, that upon the motion and touch of the one ball, motion always follows in the other. In whatever shape I turn this matter, and however I examine it, I can find nothing farther.

Abstract 9; SBN 649–50

Next, Hume turns his attention to the putative necessary connection among objects/events. He first tackles the metaphysical maxim, "*whatever begins to exist, must have a cause of existence*" (T 1.3.3.1; SBN 78–9). This is not an intuitively justifiable dictum. The two ideas, "a beginning to exist" and "a cause of existence" are not equal, and we can conceive the one without conceiving the other (for the logic of the argument, see Busch 2016: 92). Cause and effect denote objects/events that are categorically separate: the effect is not included in the cause in any way. All simple ideas are distinct, so ideas of causes and effects are distinct. We can conceive an object to not exist at one moment, then to exist at the next moment, without invoking any kind of causal principle that ties the two together by a force of necessity (T 1.3.3.3, SBN 79–80).

The upshot of Hume's inquiry is that there is no sensory evidence for any causally efficacious factor in the objects. We do not perceive a causal power. What we perceive in objects are sensory qualities. Only the quality of extension is available to us. This is notably what our senses of vision and touch tell us (T 1.2.3.4, SBN 34). Examining individual objects does not inform us about, or give us any reason to believe in, causal connections:

> There is no object, which implies the existence of any other if we consider these objects in themselves, and never look beyond the ideas which we form of them. Such an inference wou'd amount to [intuitive, demonstrative] knowledge, and

wou'd imply the absolute contradiction and impossibility of conceiving any thing different. But as all distinct ideas are separable, 'tis evident there can be no impossibility of that kind.

T 1.3.6.1, SBN 86–7 (My additions in the square brackets.)

Sensory perception informs us of the qualities of objects, but not whether they are causally related. For information about causal relations, experience is the only source. We could not make the inference from the qualities of one object to those of another without it. In Hume's account, experience is the memory and observation of constant conjunctions of species of objects/events (T 1.3.6.2, SBN 87). One perception of flame and another of heat do not suffice to establish a causal relation among the two. There needs to be repetition (how many instances of repetition, Hume does not say). Also, the observer needs to have an adequate memory. If a person, previously acquainted with the qualities of snow, were to suddenly lose her memory, she would not, by just seeing snow, be able to infer its coldness. She would just see white powder. To conclude that snow is cold, she would have to touch it several times and remember the constant conjunction of the object and its related property.[3]

The anticipation of the effect from a known cause is rooted in our habitual modes of inference, customary awareness, and natural instincts (EHU 5.5–6; SBN 43–5 and EHU 5.8; SBN 46–7). Projectivism gains further support from Hume's pronounced ambition of the *Treatise*. He establishes a new science of human nature. This is well within the limits of Hume's methodology. Dan Kervick calls Hume's approach methodological phenomenalism. He believes that Hume's method brackets physical reality altogether. Hume's humanistic science deals with our perceptions, their origins in sensory impressions, and the causal relations between ideas, and no more. Kervick allows that Hume, of course, has beliefs about the physical world, but "the study of such causes is manifestly and declaredly not the subject of the *Treatise*." For Kervick (2018: 7), "his subject is only the mental causation that occurs in the world of perceptions, by which some sensory impressions cause ideas, ideas cause other ideas, and sensory impressions and ideas cause passions and ideas of passions."

Such a methodologically phenomenalist reading is consistent with Hume's explicit objectives in his *Treatise*. For one thing, there are some passages which seem to indicate that Hume privileges his own human science over natural philosophy. In the Introduction, he puts the point as follows:

'Tis evident, that all the sciences have a relation, greater or less, to human nature; and that however wide any of them may seem to run from it, they still return back by one passage or another. Even *Mathematics, Natural Philosophy, and Natural Religion,* are in some measure dependent on the science of Man; since they lie under the cognizance of men, and are judged of by their powers and faculties.

Boehm (2013a: 59) reads this chapter as suggesting that Hume is committed to a strong form of dependence argument: all fields of inquiry are dependent on his science of human nature, including natural philosophy. If Hume restricts himself to science of human nature, and eschews natural philosophy, then causation cannot be anything more than a projection. According to a reading like this,[4] causation is not a discovery in nature, something physical out there, but a mental disposition of our human natures.

I think this line of thought is mistaken. This is not to say that causation is something fundamental, deep down. Instead, Hume thinks constant conjunctions are discovered in our physical environment. And this is all there is to discover. This is backed up by a massive amount of textual evidence. He thinks the following causal relations are discoveries:

- The dissolubility of gold in king's water (T 1.1.6.2, SBN 16).
- The law of conservation of momentum (EHU 4.13; SBN 31).
- Water potentially causes drowning to non-aquatic beings (EHU 4.6; SBN 27).
- Flame causes heat (EHU 4.6; SBN 27).
- Two pieces of marble are easily separated by lateral force, instead of exerting force in a direct line (EHU 4.7; SBN 28).
- Gunpowder is explosive (EHU 4.7; SBN 28).
- Lodestone is attractive (EHU 4.7; SBN 28).
- Collision of billiard balls produces motion (EHU 4.9; SBN 29).
- Objects fall to the ground by the force of gravity (EHU 4.9; SBN 29).
- Crystal is a product of very high temperature (EHU 4.13; SBN 31–2).
- Ice is a product of cold temperature (EHU 4.13; SBN 31–2).

On this list all, except one item, are mentioned in the first *Enquiry*. Each proposition concerns matters of fact. Factual propositions concern "real existence," not only ideas in the mind, as in the case of relations of ideas. They convey information about the world around us. As has been already established, factual reasoning is founded on the relation of cause and effect: "All reasonings concerning matter of fact seem to be founded on the relation of *Cause and Effect*. By means of that relation alone we can go beyond the evidence of our memory and senses" (EHU 4.4; SBN 26–7). We acquire the ideas of the qualities of objects via sensory perception. Mere perception is not enough for factual claims about anything. Even such a mundane example as "coffee is hot" is based on memory and frequent observation of the conjunction of skin/lips and liquid. Causal reasoning cannot be grounded in perceptions alone; it is founded on experience: "I shall venture to affirm, as a general proposition, which admits of no exception, that the knowledge of this relation is not, in any instance, attained by reasonings *à priori*; but arises entirely from experience, when we find, that any particular objects are constantly conjoined with each other" (EHU 4.6; SBN 27).

In the next paragraph, Hume is explicit about the discovery of causation by experience: "*causes and effects are discoverable, not by reason, but by experience*" (EHU 4.7; SBN 28).[5] He comes up with three examples of physical causation, involving the behavior of marble, gunpowder, and lodestone. A person without experience cannot know that two pieces of marble can be easily separated by lateral force, instead of direct force. Likewise, the person could not know that gunpowder explodes if it is set on fire, or that magnetic items attract metals in their vicinity. When Hume argues that a person cannot discover these causal relations by relying on pure reason, I think he means that the regular relations between the objects/events are already there before the person discovered them. To show this, we should look carefully at the negative form of Hume's argument. First Hume says that we are utterly unable to foretell what effects some causes may have. Then he gives the example of the separation of pieces of marble: "Present two smooth pieces of marble to a man, who has no tincture of natural philosophy; he will never discover, that they will adhere together, in such a manner as to require great force to separate them in a direct line, while they make so small a resistance to a lateral pressure" (EHU 4.7; SBN 28).

A person without the relevant natural scientific knowledge does not know, before experimenting with the pieces of marble, the right answer. For the question—Are the pieces more easily separated by a lateral or a direct force?—has a right and a wrong answer. Before experimenting with the materials, getting the right answer is a matter of sheer luck. After satisfactory testing, one option is falsified and the other corroborated. As there is a right answer, a discovery, there needs to be something in the world that makes it true.

Hume does not himself use the term "truth-maker." This comes from the work of D. M. Armstrong and other contemporary metaphysicians (the basic idea is Aristotle's). Truth-maker metaphysics tries to answer the following question: "What is there in the world in virtue of which these truths are true?" (Armstrong 2004: 3). I think such a question about truth-makers is applicable to Hume's philosophy of causation. In his search for the idea of cause, Hume could not find the idea of a causal power. But he did find contiguity, succession, and constant conjunction. In other words, he found regularity. It is therefore the regular, contiguous, and temporally linear conjunctions among objects/events that are there, in the world. Our environment is rife with regularities. The following line of reasoning in the *Treatise* (T 1.3.14.26–8; SBN 167–9) summarizes my reading of Hume's conception of causality. He starts first by distancing himself from projectionism. Then he goes on to deny the view of necessary connection among causally related objects:

> What! the efficacy of causes lie in the determination of the mind! As if causes did not operate entirely independent of the mind, and wou'd not continue their operation, even tho' there was no mind existent to contemplate them, or reason concerning them. Thought may well depend on causes for its operation, but not causes on thought. This is to reverse the order of nature, and make that secondary, which is really primary [...] As to what may be said, that the operations of nature are independent of our thought and reasoning, I allow it [...] But if we go any farther, and ascribe a power or necessary connexion to these objects; that is what we can never observe in them, but must draw the idea of it from what we feel internally contemplating them.

What is discovered is constant conjunction, not necessary connection. Hume (EHU 7.28; SBN 75–6) explains this by using his paradigm example, the collision of billiard balls: "The first time a man saw the communication of motion by impulse, as by the shock of two billiard-balls, he could not pronounce that the one event was *connected:* but only that it was *conjoined* with the other." Here Hume points out that we do not immediately perceive any connection among causally related objects/events. Rather, we perceive a conjunction of two discrete objects, the billiard balls. This is not to say that we lack any idea of a connection. Certainly, we do have an impression of necessary connection, but this requires multiple experiences of conjunctions. These repeated experiences are needed to form an internal impression of reflection in the mind. "This connexion, therefore, which we *feel* in the mind," is a "customary transition of the imagination from one object to its usual attendant, is the sentiment or impression, from which we form the idea of power or necessary connexion. Nothing farther is in the case" (ibid.; see also Millican 2009: 647–8).

Don Garrett (2015a: 74) explains this in a way that is congenial to my reading: "The impression of necessary connection cannot be the sensory perception of something external to the mind and located in the cause and effect; instead it must be something new that arises in the mind itself as the result of the repeated conjunction." The source of the information we have of the relation of cause and effect is experience, which Hume understands as being observation and memory of constant conjunction among species of objects or events (T 1.3.6.2; SBN 87). The knowledge we have of objects or events is provided by perceptions according to the copy principle. Thus factual discourse in natural philosophy is limited to perceptions of objects or events and their constant conjunctions. "School metaphysics" and "divinity" fall outside the scope of perception and experience, so they may be committed to the flames, as they "contain nothing but sophistry and illusion" (EHU 12.34; SBN 165).

So far, I have argued that Hume i) subscribes to a bottom-up analysis of laws, ii) thinks that laws are causal, and iii) his concept of causation in natural philosophy maintains that causation is discovered constant conjunction. I have not yet scrutinized Hume's actual references to the laws of physics. Next, I wish to take a closer look at Hume's analysis of various laws of motion. I propose that there are two things to note. First, Hume does not eschew mechanical philosophy altogether: his concept of causation tacitly assumes a mechanism. This is apparent in his rules for causal reasoning. Nevertheless, as I shall also argue, he remains agnostic on whether bodies' internal constitution is machine-like. This agnosticism extends to hidden causal powers that bring about the phenomena of motion.

Mechanism and Causal Power in Laws of Nature

In my interpretation, Hume explicitly recognizes (at least) five laws of physics. Before analyzing them further, let me issue a word of caution. He does not present laws in an axiomatic form. As I pointed out in Chapter 2 of this book, this is because he is not first and foremost a natural philosopher. His ambition is to provide a mental geography by mapping the cognitive structures of the human mind. For this purpose, laws of nature are irrelevant. Hume does not have a settled physical theory that he articulates throughout all his works. Instead, on various scattered occasions, he refers to a number of laws of motion in the process of making other points relevant to his overall philosophical project. These references do not explicitly tell us whether Hume approves or disapproves of the propositions in question. He might quote from natural philosophers or attribute them to common viewpoints. He does not seem to have any deep concerns about their consistency. But Hume's references to laws of nature do tell us, or so my argument propounds, about his tacit mechanistic assumptions wedded to his notion of causality (which is not the same as a commitment to mechanism as the causally efficacious factor in causation). In various unrelated contexts, Hume expresses physical laws as follows:[6]

- A body at rest or in motion continues for ever in its present state, till put from it by some new cause (EHU 7.25n; SBN 73n16).
- A body impelled takes as much motion from the impelling body as it acquires itself (EHU 7.25n; SBN 73n16).
- The moment or force of any body in motion is in the compound ratio or proportion of its solid contents and its velocity (EHU 4.13; SBN 31).
- The equality of action and re-action seems to be a universal law of Nature (DNR 8.11; 186).
- Gravitation of matter . . . produces a motion from the one to the other (T 2.3.8.8; SBN 434–35).

The fact that Hume mentions the first law is neither surprising nor controversial. The law is the principle of inertia and can be found both in Descartes' and Newton's *Principias*. In Descartes' formulation, the law states: "The first law of nature: that each thing, as far as is in its power, always remains in the same state; and that consequently, when it is once moved, it always continues to move" (Pr II 37). In his second law he adds that "all movement is, of itself, along straight lines" (Pr II 39). Newton combined the two laws into one (Slowik 2017: Section 4): "Every body perseveres in its state of being at rest or of moving uniformly straight forward, except insofar as it is compelled to change its state by forces impressed" (*Principia*, Axioms, or the Laws of Motion). Hume discusses this law in the context where he declines Malebranchean theistic occasionalism. In the Malebranchean view, God is the only causal agent in the world. Malebranche (1977) thought, to quote from Steven Nadler (2000: 115–16), that "God is directly, immediately, and solely responsible for bringing about all phenomena." Whether we are dealing with bodily, mental, or mind–body causality, these are all occasions for God to cause an act of His will in as much as it is in accordance with the laws of nature (except for miracles). In Hume's account, we know inductively that there are certain laws of motion. But we are not justified to infer that God constantly recreates the world to prevent the dissipation of forces. Hume is simply unwilling to discuss the theological and metaphysical underpinnings of *vis inertiae* or *vis viva*.

The second law on the list is a conservation law. Hume mentions this in the same footnote of his first *Enquiry* as the first law, but he also analyzes a definition used in the law. The second law applies in the special case when the masses of two bodies of an isolated system are equal. After the impact of say, two billiard balls, the velocity of the object ball will be the same as the initial speed of the cue ball. The third law is a definition which explicates the constant, the preserved quantity in law number two. Hume mentions the third law (to be precise, the third is not a law but a definition used in law two) when discussing the notion of mixed mathematics. Law three says that the total momentum or force (Hume conflates the two) equals the bulk of the matter and the resultant velocity of the body jointly. Hume goes on to explain the definition in this law further. He writes that "consequently . . . a small force may remove the greatest obstacle or raise the greatest weight, if, by any contrivance or machinery, we can encrease the velocity of that force, so as to make it an overmatch for its antagonist" (EHU 4.13; SBN 31). This quotation indicates that Hume thinks that force is proportional to the product of the weight and the speed of a body. Hume may have got this conception from Huygens. It might be an imitation of his work which equated forces with the powers of "raising a weight" (Gabbey 1980).[7]

Hume's fourth law in the list is Newton's third law in his *Principia*. Hume mentions it only in the *Dialogues* (8.11). It is part of Philo's argument that challenges a received view according to which a divine mind can create matter. This argument points out an inconsistency in the view that thought might influence matter. Because of Newton's third law, it was known that action on "matter" has "an equal reciprocal influence upon it" (DNR 8.11; 186). If non-material mind affects matter, the matter should affect the non-material mind with equal and opposite force. But we do not see this happening. The received view which maintains that an immaterial mind causes changes in the material world is hence inconsistent with the known laws of nature. Newton himself puts his third law as follows: "To any action there is always an opposite and equal reaction; in other words, the actions of two bodies upon each other are always equal and always opposite in direction" (*Principia*, Axioms, or the Laws of Motion). On a superficial level, Hume agrees with Newton's formulation. But there is a crucial aspect missing in Hume's understanding. This lack of understanding is relevant in assessing the relation between Newton's third law and the concept of force. Although Philo's line of argument in the *Dialogues* (8.11) brings up the notion of interaction, Hume does not consistently understand the concept of force in terms of interaction. To expound on this difference, it is useful to quote from Max Jammer's (1957: 127) study of the history of the concept of force:

> The third law ... supplies an additional important characteristic of force not
> mentioned previously [in Newton's first and second laws of motion]: force

manifests itself invariably in a dual aspect; it is action and reaction simultaneously. Much as business transaction can be regarded both as a purchase and as a sale of the same amount, force can be considered as action as well as reaction of the same magnitude.

In Newton's account, if there were just one particle of mass in the universe, no forces would appear.[8] Forces are not qualities hidden in bodies but dynamic relations among them. When Hume refers to the concept of force, he seems to understand it as a property of an individual object (interestingly, this also seems to have been Newton's position in his pre-*Principia* tract "De Gravitatione").[9] Hume notes that force is "the unknown circumstance of an object" (EHU 7.29; SBN 77n17). Force fixes and determines "the degree or quantity of its effect [the resultant motion of a body]." The concept of force is an instrument which enables one to predict the change of motion of bodies. Accordingly, Hume does not emphasize the interactive dynamic character of forces. He does not say that forces appear among particles of mass, or that mass is the resisting factor in acceleration.

On first reading, the last, fifth, law Hume mentions seems to be a rather generic description of gravitation in a two-body system. We might think that Hume is quoting from Newton's *Principia*, as its first edition is the original publication for the argument of the law of universal gravitation. In its original context, Hume is not however leaning on Newton. Hume uses the fifth law as an example of a body moving from a higher spatial position to a lower one due to gravitation: "There is no natural nor essential difference betwixt high and low" because "this distinction arises only from the gravitation of matter, which produces a motion from the one to the other" (T 2.3.8.8; SBN 434–5). Hume's depiction here is clearly not a reference to Newton's law of universal gravitation. But in the second *Enquiry* (EPM 6.6; SBN 236), Hume mentions a particular case of Newton's law of universal gravitation, as he claims that the moon is kept "in its orbit by the same force of gravity, that makes bodies fall near the surface of the earth." This makes his understanding of gravitation somewhat Newtonian. There are still the following restrictions: Hume does not mention the direct proportionality of masses or the inverse-square proportionality of distance in gravitational attraction.

As can be seen in the previous analyses of various laws of nature, Hume has no doubt that laws are causal. There is a cause and effect pair discoverable in laws of nature. In his view, forces and momenta cause change of motion. He also seems to be sympathetic in some way to mechanical philosophy. His references to physical laws indicate a more Cartesian-mechanistic understanding of laws, rather than a Newtonian dynamic one. What makes Hume's conception a distinctly mechanical one are the explicit constraints on causation. These constraints are expressed in his rules by which to judge causes and effects (T 1.3.15; SBN 173).

There are many regularities that are not causal relations. Night follows day, a storm follows a drop of a mercury column in a barometer, and not getting pregnant follows a woman taking contraceptive pills. Hume recognizes that this is an obvious difficulty in his regularity account: "Where objects are not contrary, nothing hinders them from having that constant conjunction, on which the relation of cause and effect totally depends" (T 1.3.15.1; SBN 173). "Since therefore 'tis possible," he continues, "for all objects to become causes or effects to each other, it may be proper to fix some general rules, by which we may know when they really are so" (T 1.3.15.2; SBN 173). The result is eight rules by which to judge causes and effects. Hume formulates five conditions that need to be obtained between causes and effects: 1) contiguity, 2) succession, 3) constancy, 4) sameness, and 5) commonality. The last three rules are more clarifications. 6) states that, at first sight, an effect might follow from an uncommon cause. Under closer scrutiny, the unexpected effect is revealed to be the result of a different cause. 7) assesses the proportional nature of causal relations. The degree of the cause might be diminished or augmented to a certain degree, but not beyond it, or otherwise it will lead to a different cause. 8) remarks that an object which exists for some time is not in itself the cause for some effect; if a cause does not produce its effect immediately, it is only a partial cause (Garrett 2018: 63). Without contiguity in time and place, as well as constant conjunction, an object cannot be said to be a cause for some effect.

As Eric Schliesser (2007: Section 4.5) notes, the source of the eight rules is unclear. Hume gives some hints of their origin. They "might have been supply'd by the natural principles of our understanding" (T 1.3.15.11; SBN 175). Commenting on his fourth rule, Hume notes that "this principle we derive from experience, and is the source of most of our philosophical reasonings" (T 1.3.15.6; SBN 173–4). This suggests that, on the one hand, our judgments concerning causal relations have perfectly natural and empirical origins. On the other hand, Hume also emphasizes the restricting and regulative nature of his rules: "We shall afterwards take notice of some general rules, by which we ought to *regulate* our judgment concerning causes and effects; and these rules are form'd on the nature of our understanding, and on our experience of its operations in the judgments we form concerning objects" (T 1.3.13.11; SBN 149; see also De Pierris 2001 and Martin 1993).

I shall focus on two of his criteria: contiguity and the separability of causes and effects. These criteria imply that Hume is, in a relevant sense, a mechanical philosopher. To advance an argument about Hume's assumption of mechanism in the laws of nature, it is important to clarify the notion of "mechanical philosophy." This is not a settled term. It cannot be defined by listing a few necessary and sufficient conditions which should be included in the definition. Rather, "mechanical philosophy" is an umbrella term which has various interrelated meanings. In the formulation of Sophie Roux and Daniel Garber (2013: xi), it can be interpreted to stand for the following commitments:

- "the comparison of natural phenomena, most specifically the world and animals, to existing or imaginary machines";
- "the general program of substituting for the 'common philosophy', i.e. the scholastic philosophy, a new philosophy, still to be identified";
- "the more specific rejection of Aristotelian hylomorphism and the correlated adoption of an ontology according to which all natural phenomena can be understood in terms of the matter and motion of the small corpuscles that make up the gross bodies of everyday experience alone."

Hume's concept of causation instantiates mechanical philosophy in the first two senses. Hume's criterion of contiguity indicates that he formulated his causal philosophy as taking its model from the workings of machines. His argument for the separability of causes and effects indicates his rejection of the older Aristotelian philosophy of causation in which "effects" are included in "causes," hence corresponding to point two. Nevertheless, Hume is not a mechanical philosopher in the sense of the third point. His radically empiricist and skeptical theory of perception and causation does not license us to infer that bodies have a machine-like microstructure, as the corpuscularian hypothesis suggests.

Contiguity

Newtonian dynamics violate a core aspect of mechanical philosophy. Newton's third law together with the law of universal gravitation imply instantaneous action at a distance among all the particles of mass in the universe. No matter how long the distance, or how small the masses, the laws countenance instant nonmediated causal action between the particles. Newton understands space as being an empty[10] Boylean vacuum, a place that bodies fill. There is nothing in space which "might impede, assist, or in any way change the motions of bodies" (*Principia*, General Scholium, "De Gravitatione," Definitions 2 and 4). Thus there is instantaneous action among discrete objects, although there is nothing in between them. In his *Treatise*, there is one central paragraph (T 1.3.2.6; SBN 75) in which Hume discusses contiguity in physical causation. This implies his suspicion of action at a distance:

> I find in the first place, that whatever objects are consider'd as causes or effects, are *contiguous*; and that nothing can operate in a time or place, which is ever so little remov'd from those of its existence. Tho' distant objects may sometimes seem productive of each other, they are commonly found upon examination to be link'd by a chain of causes, which are contiguous among themselves, and to the distant objects; and when in any particular instance we cannot discover this connexion, we still presume it to exist. We may therefore consider the relation of contiguity as essential to that of causation.

Hume suggests that it is desirable to find an explanation which includes a reference to contiguity in dynamic laws. If contiguity is not discovered in causal relations, it is still presumed to exist. Hume posits that in physical causation there needs to be contact among causally related bodies. In this respect, Hume lends his support to Cartesian cosmology.[11] Both Hume and Descartes are very hesitant in accepting the existence of a vacuum. There is no empty space independent from bodies (or at least, we have no clear and distinct idea of it). Descartes claims in his *Principles* (II 16) "that it is contradictory for a vacuum, or a space in which there is absolutely nothing, to exist." In the same manner, Hume claims that it is "impossible to conceive either a vacuum and extension without matter" (T 1.2.4; SBN 40).[12]

As Hume approves this part of Cartesian cosmology, there is no action at a distance across empty space because there is no empty space in the first place (or we do not have its putative idea).[13] In this mechanistic model of causation, "phenomena result," Katherine Dunlop (2012: 86) writes, "from something like pushing or pulling, localized to the surface of the body." Hume is a mechanical philosopher in this sense. Bodies' motions are generated in a way that is reminiscent of the way mechanical devices produce a chain of motion. This is apparent in the way that, for example, a water mill is used to crush grains. There is a succession of physical contacts among the parts of the machine.

In the first *Enquiry*, the contiguity requirement disappears. Its absence could be interpreted as Hume's shift from Cartesianism to Newtonianism. This might be in part true; the *Enquiries* are maybe somewhat more Newtonian than the *Treatise*. Another reason for Hume to drop contiguity might have been the recognition, stated explicitly already in the *Treatise*, that many entities that do not exist in space can enter into causal relations. The initial discussion on contiguity in *Treatise* 1.3.2.6 includes a reference to T 1.4.5, probably to its paragraphs 9–12. In paragraph 10 Hume provides a metaphysical maxim according to which "an object may exist, and yet be nowhere" (T 1.4.5.10; SBN 235). He finds it very plausible that "the greatest part of beings do and must exist after this manner." There can be constant conjunctions between objects of perception "which exists without any particular place" (T 1.4.5.12; SBN 237), and these conjunctions still count as causal relations. We are only left with speculation as to why Hume does not explicitly mention contiguity in his treatment of causation in the *Enquiries*. But I do not think that he completely gave up his initial suspicion of action at a distance. A perusal of the first *Enquiry*—with respect to his concept of causation, at least—suggests that he does not dismiss mechanical philosophy altogether. Hume's assumption of mechanical philosophy is apparent in many of the examples he provides on causality: collision of billiard balls, vibrations of strings, operations of clocks, strings, wheels and pendulums (EHU 4.9, 7.29, 8.13; SBN 29, 77, 87). These examples indicate that he is not, even in the first *Enquiry*, entirely jettisoning the mechanical model of causation. Moreover, Hume does not make the positive claim that there is causal action at a distance.[14]

Separability of Causes and Effects

In the first *Enquiry*, Hume claims explicitly that cause and effect are distinctly separable: "In a word, then, every effect is a distinct event from its cause" (EHU 4.11; SBN 30). This claim suggests that Hume is a mechanical philosopher in the sense that he diverges from the preceding Aristotelian tradition in natural philosophy. In the Aristotelian framework, various parts of a causal process are not distinctly separable. An acorn is potentially a tree. In its essence, an acorn strives to grow to be a tree. The final cause of the process, the resultant tree, is included in the potential cause, the acorn. According to the mechanical philosophy, objects change their state by external causes acting upon them. This is, in the broadest sense, the content of Descartes' first law: If there is to be any change in an object, there needs to be an *external* cause for it (Brading 2012: 13, fn. 1). In this mechanistic model of causation "it would be a category mistake," Ott (2009: 14) argues, "to think that an event, or a mode of a body, could include its effect."

Hume's argument for the distinctness of causes and effects follows from his separability principle. Briefly put, the principle maintains that separability implies distinctness. Hume writes: "Whatever objects are separable are also distinguishable, and that whatever objects are distinguishable are also different" (T 1.1.7.3; SBN 18). Cause and effect stand for particular objects or events. They are separate from each other because one can be conceived without conceiving the other. There is no contradiction in conceiving a cause and not conceiving an effect: We can "conceive any effect to follow from any cause," Hume asserts (T Abstract 11; SBN 650, see also T 1.3.14.13; SBN 161–2, and EHU 4.9–11; SBN 29–30). In Hume's account, the mind has a threshold in forming ideas: there needs to be a minimum sensible, to wit, a simple impression. For example, I now perceive a white office desk and a red coffee mug in front of me. My visual perceptions of these objects evince that they are distinct. Their different colors enable me to decide that there are two different finite objects with distinct spatial boundaries. This does not of course explain why the objects are separate. This only raises the question of why the impressions causing the ideas are separate. However, this reasoning illustrates the point of why effects are not included in causes. My visual perceptions do not inform me whether the objects stand in a causal relation or not. In the first *Enquiry* (EHU 4.6; SBN 29), we find a lengthy argument for refusing to include effects in their causes in physical causation. The example Hume uses below—motion produced in the collision of billiard balls—is very typical for a mechanical philosopher:

> The mind can never possibly find the effect in the supposed cause, by the most accurate scrutiny and examination. For the effect is totally different from the cause, and consequently can never be discovered in it. Motion in the second Billiard-ball is a quite distinct event from motion in the first; nor is there any thing in the one to suggest the smallest hint of the other.

Because of Hume's separability principle, his position is mechanistic also in the sense that it is incompatible with the notion of interaction in dynamics. Consider the following scenario explained by Newton's third law of motion. I press the table with my hand; the table presses my hand with equal and opposite force. What is the cause, and what is the effect in this scenario? Is my pressing of the table the cause, or is the pressing coming from the table? In Newtonian dynamics forces are generated through interactions. Although Newton's second law is the causal law and his third law is rather a law of co-existence (Tooley 2004: 88), it is still difficult to separate the supposed cause and the supposed effect in dynamic interactions (as forces appear between mass points). But this is what Hume's separability principle requires.

Laws of nature are causal for Hume, and his concept of causation—as can be seen from his references to various laws—assume mechanistic collisions between bodies. But what is the causally efficacious factor of bodies? Locke and Boyle think it is the causal power, or mechanism of bodies. Hume thinks, as I will argue, that we should remain agnostic about such causally efficacious, mechanistic explanations. Explicating the difference between these scholars shows that although Hume is willing to accept some tenets of the mechanical philosophy, he does not go so far as to posit powers to explain manifest regularities.

Causal Power

To reiterate a part of the definition of Roux and Garber (2013: xi), mechanical philosophy adopts "an ontology according to which all natural phenomena can be understood in terms of the matter and motion of the small corpuscles that make up the gross bodies of everyday experience alone." Ephraim Chambers' (1728: 521) contemporaneous definition is similar to Roux's and Garber's: "Mechanical *Philosophy*, is the same with the Corpuscular Philosophy; viz. that which explains the Effects of Nature, and the Operations of Corporeal Things, on the Principles of *Mechanics*; the Figure, Arrangement, Disposition, Motion, Greatness or Smallness of the Parts which compose natural Bodies. See Corpuscular."

Boyle and Locke are mechanical philosophers in the sense provided by the above definitions. In his "Grounds for and Excellence of the Corpuscular or Mechanical Philosophy," Boyle (1989: 117) argues that the size, shape, and motion of corpuscles, which make the "texture" (microstructure) of matter, determine all macroscopic corporeal phenomena. This "texture," for Boyle, is the structure that is "made out of minute and insensible corpuscles" of bodies. Locke is very sympathetic to Boyle on this issue. In explaining the texture of bodies, and how it relates to experiential macroscopic causation among bodies, both Boyle (*Origin of Forms and Qualities*, section 2) and Locke (*Essay* IV.iii.25) compare it to human-made machinery. Locke writes that

> These INSENSIBLE CORPUSCLES, being the active parts of matter, and the great instruments of nature, on which depend not only all their secondary qualities, but also most of their natural operations . . . I doubt not but if we could discover the figure, size, texture, and motion of the minute constituent parts of any two bodies, we should know without trial several of their operations one upon another . . . a smith [would] understand why the turning of one key will open a lock, and not the turning of another. But whilst we are destitute of senses acute enough to discover the minute particles of bodies, and to give us ideas of their mechanical affections, we must be content to be ignorant of their properties

and ways of operation; nor can we be assured about them any further than some few trials we make are able to reach.

Locke's explanation is founded on an argument from analogy. A locksmith can predict which one of several keys can open a lock: "a smith [would] understand why the turning of one key will open a lock, and not the turning of another" (ibid.). In the same way, an observer can predict the outcome of the collisions of macroscopic bodies. Neither the locksmith nor the observer relying on their senses has direct empirical access to the machinery of locks or bodies. Nevertheless, in both cases the observable effects ensue from the hidden internal micro-constitutions of the lock and of the bodies. For Locke, the functioning of artifactual machinery is therefore analogous to causal efficaciousness in macroscopic causal relations.

Are corpuscles, or the mechanism of body, causally efficacious in Hume's view, too? I think Hume has to be agnostic on this matter. There is no evidence coming from our five senses to confirm the existence of corpuscles. We do not perceive how these putative microscopic parts interact. Do they squeeze, enlarge, vibrate? And how do these hypothetical movements change the motion of bodies in collisions? If two objects or events stand in causal relations, we should perceive both objects and events.[15] But we only see motions of objects, and their collisions. Hume is very clear on this point in the Abstract (7–10; SBN 648–50) to the *Treatise*. First he recounts his copy principle by pointing out that all meaningful terms must be annexed to impression-based ideas. Then he examines what we perceive in a paradigm case of causation, that is, the collision of billiard balls, "when both the cause and effect are present to the senses":

> The first ball is in motion; touches the second; immediately the second is in motion: and when I try the experiment with the same or like balls, in the same or like circumstances, I find, that upon the motion and touch of the one ball, motion always follows in the other. In whatever shape I turn this matter, and however I examine it, I can find nothing farther.

Hume maintains that we experience a conjunction among objects and events, but not any connection (EHU 7.21; SBN 69–71):

> When we look about us towards external objects, and consider the operation of causes, we are never able, in a single instance, to discover any power or necessary connexion; any quality, which binds the effect to the cause, and renders the one an infallible consequence of the other. We only find, that the one does actually, in fact, follow the other. The impulse of one billiard-ball is attended with motion in the second.
>
> EHU 7.6; SBN 63

It is true that Hume entertains the idea of a causally efficacious power. He uses causal talk when addressing "the power or force, which actuates the whole machine [of the universe]," but which is still "entirely concealed from us, and never discovers itself in any of the sensible qualities of body" (EHU 7.8; SBN 63–4 [My additions in the square brackets.]). This quote strongly resembles Locke's corresponding views about causal power. Nevertheless, after such surmising, Hume goes on to clarify his position. In the very next paragraph (EHU 7.9; SBN 64), he asks whether the idea of necessary connection is "derived from reflection on the operations of our own minds, and be copied from any internal impression." As we experience constant conjunctions over and over again, we simply form a customary feeling in the mind about causal necessity. Such necessity is an impression of reflection. In the same way as an idea of a delicious food causes an impression of desire, an idea of constant conjunction of objects or events causes an impression of necessity. We feel *as if* things were related by necessity and governed by a causal power in the physical world.

Given Hume's insistence that causation is identified with experience of constant conjunctions of objects/events, and that we should perceive these objects/events by means of the copy principle, we should also remain agnostic on whether bodies are made of corpuscles, and whether this putative micro-constitution of matter determines macroscopic causation. However, my interpretation does not contend that on Hume's account we could not have the idea of a corpuscle (or that the copy principle makes the term "corpuscle" a meaningless one). Considering the traditional, Democritean atomist doctrine (see Berryman 2016), a corpuscle is a body that has a certain shape, size, position, and orientation. These are manifest features, all of which satisfy the copy principle. The features described by classic atomist doctrine or modern corpuscularianism are observable in the macroscopic bodies around us. In principle, a corpuscle is not imperceptible, and therefore it is not inconceivable. Hume states that it "is certain, that we can form ideas, which shall be no greater than the smallest atom of the animal spirits of an insect a thousand times less than a mite" (T 1.2.1.5; SBN 28). In Hume's time there were no microscopes of adequate resolution to detect the hypothetical corpuscles,[16] so he does not find any reason to believe in the existence of these particles. This is a reservation that stems from inadequate technology; the copy principle does not rule out corpuscles.

Hume diverges from Boyle and Locke because in his rigidly empiricist and skeptical theory of perception and causation, we are not licensed to infer what the hidden micro-constitution of bodies is, or what relation this putative constitution is to macroscopic bodily causation. This doubt concerning the texture of bodies arises, to quote from Graciela De Pierris (2015: 14), from Hume's "consistent and radical interpretation of the sensible phenomenological model of ultimate evidence." Still, De Pierris' (ibid., fn. 25) interpretation accepts that corpuscles are in principle observable. She writes: "Causation enables us to make inferences from instances of constant conjunction that have been observed to those that have not yet been observed but still are, in principle, *observable*." For Hume, laws of nature are generalizations over regularities. He does not find anything that would ground these regularities, because there is no sensible evidence for that putative ground.

One critical question should be raised here. Boyle and Locke posit corpuscles to explain observable phenomena. This basic point can be framed by using a vocabulary borrowed from our contemporary philosophy of science: we should believe in the existence of corpuscles because they are an inference to the best explanation (of explaining macroscopic bodily causation). Does Hume's account of causation rule out such an inference? I think the answer is yes. An inference to the best explanation does not guarantee an inference to the right explanation. Hume does not apply abductive inference like Boyle and Locke in their argument from analogy. This is because his account of causation recognizes that in principle any thing may be the cause of another thing (T 1.3.14.13; SBN 161–2, T Abstract 11; SBN 650, and EHU 4.9–11; SBN 29–30). The only way for us to have information of any causal relation is by experience. As already noted, Hume thinks experience is memory and observation of species of objects being constantly conjoined (T 1.3.6.2; SBN 87). I remember having observed that every time I place my finger to a candle flame, I feel heat, and every time I put my hand into a bucket of ice I feel cold. Objects or events that stand for causes and effects need to be perceivable. So we know about these causal relations by experience. But we do not have the same experience when it comes to mechanist explanation of bodily causation. We do not perceive (even with the best microscopes available) the sizes, shapes, positions, or orientations of the corpuscles, and the way they relate to the collisions of average-size objects. We can only speculate on this. A speculation like this is, in Hume's parlance, a causal hypothesis which goes beyond experience and should be declined (T Intro 8; SBN xvii; compare this with Newton's experimentalism and criticism of hypotheses). Therefore Hume remains an agnostic on the existence of corpuscles, and their putative causal efficaciousness.

This interpretation explains Hume's announced dismissal of Boyle's corpuscularian theory. In his *History* (VI: 540), Hume highly appreciates Boyle's experimental work with the air pump. Still he goes on to denote the corpuscularian position of Boyle's mechanical philosophy "imaginary":

> Boyle improved the pneumatic engine invented by Otto Guericke, and was thereby enabled to make several new and curious experiments on the air as well as on other bodies: His chemistry is much admired by those who are acquainted with that art: His hydrostatics contain a greater mixture of reasoning and invention with experiment than any other of his works; but his reasoning is still remote from that boldness and temerity, which had led astray so many philosophers. Boyle was a great partizan of the mechanical philosophy; a theory, which, by discovering some of the secrets of nature, and allowing us to imagine the rest, is so agreeable to the natural vanity and curiosity of men.

The nuance of the preceding quote is mostly positive. Hume reveres Boyle's experimentalism, and even partly his mechanical philosophy by contending that mechanical philosophy can discover "some of the secrets of nature." This is consistent with the fact that Hume is both a supporter of the British experimental tradition in natural philosophy and, in a relevant sense, a mechanical philosopher. His disagreement with Boyle shows the limits of his mechanical philosophy: Hume does not believe in Boyle's minute posits, the corpuscles. There is further textual evidence for this interpretation. In the next paragraph of *History* (VI: 542), Hume writes: "While Newton seemed to draw off the veil from some of the mysteries of nature, he shewed at the same time the imperfections of the mechanical philosophy; and thereby restored her ultimate secrets to that obscurity, in which they ever did and ever will remain."

Eric Schliesser (2007: Section 4.2) reads the above quote as indicating "Hume's treatment of Boyle reveals that he thought it was a good thing Newton falsified the mechanical philosophy." But, he adds: "Hume acknowledges that the mechanical philosophy could offer some successful explanations." Schliesser's reading is consistent with my position: Hume discredits corpuscularianism *and* simultaneously thinks that mechanical philosophy can in part provide cogent explanations of natural phenomena.

In this chapter, I have argued that Hume's copy principle and concept of causation make him an agnostic concerning causal power. This is consistent with the traditional reading of Hume on causation. For Hume causation is discovered constant conjunction. However, we do not acquire the idea of a necessitating causal power from the causal relations themselves. Hume's concept of causation is reminiscent of the Cartesian-mechanistic tradition in natural philosophy, but it is not committed to the claim that causes necessitate their effects, or that laws of nature cause things to happen. Next, I will argue that Hume does not eschew forces in physics altogether like Berkeley. Forces are mathematical instruments that make predictions possible. We lack the ideas of causally efficacious parts of bodies; but we do clearly understand the mathematical proportions of forces and momenta.

Force as a Calculating Device for Predictions

Although I have argued that causal relations are discovered for Hume, we do not have the idea of causal power itself, located in the hidden configuration of bodies. We should have a clear image (the idea) of the mechanism, which we do not have. But we clearly understand the mathematical dimensions of forces. To shed light on Hume's position, it is worth analyzing Berkeley's views on this matter first.

In his *De Motu*, Berkeley (1992) seems to recommend a rather strong instrumentalist interpretation of Newtonian dynamics. According to Lisa Downing's (1995: 200) seminal reading, positing forces, which are supposed to be "corporeal, that is, physical qualities" is in contradiction with the supposition that forces would also be "active [...] efficient causes of motion." Berkeley understands physical bodies to be inactive. They cannot originate their motion by themselves. A rock held in the hand cannot willingly decide to fall to the ground nor can a planet willingly decide to gravitate toward other planets. If it were the case that bodies contain some intrinsic powers which contribute to their motion, they would have to be, Berkeley thinks, some kind of animate and ensouled living creatures (Downing 2005: 239–40). But brute matter cannot think; it cannot decide to go toward another piece of matter whose spatial location it knows. That would be truly occult. To eschew the problem, Berkeley is leaning toward scientific instrumentalism, thus avoiding the awkwardness that he sees to be inherent in the realist interpretation of forces.

The structure of Berkeley's argument against realism on forces, as Downing (1995: 201; 2005: 246–7) reformulates it, is as follows. Realist outlook supposes forces "to be active (i.e. causally efficacious) qualities of body." From Berkeley's epistemological point of view, "all the known qualities of body are passive." We cannot infer the cause of motion from the empirical attributes of bodies. Properties like "impenetrability, extension, and figure" (DM 22), the tangible or visual features of bodies, cannot be felt or seen to originate motion. The principle, which is supposed to produce motion, which is supposed to be the *cause* of motion, is hidden from us.

When we observe a free fall of a body, "we," Berkeley explains, "perceive in descending heavy bodies an accelerated motion toward the center of the earth: but beyond this we perceive nothing by sense" (DM 4). Gravity is not a quality that can be detected by sense experience: "By reason however we gather that there is some cause or principle of these phenomena, and this is commonly called 'gravity'. But since the cause of the descent of heavy bodies is unseen and unknown, gravity in this sense cannot properly be called a sensible quality. It is therefore an occult quality" (DM 4).

Berkeley holds realism in dynamics to be problematic in semantic aspects, too. Since "force is an unknown quality of bodies," "the term 'force' is empty" (Downing 1995: 204). Berkeley contends that forces are unobservable entities whose meaning we do not properly understand: "From what has been said it is manifest that those who affirm that active force, action, and the principle of motion are truly in bodies embrace an opinion based on no experience, and they add to it with obscure and very general terms, nor do they adequately understand what they themselves mean" (DM 31).

Berkeley interprets Newton's laws as calculating devices. As Berkeley writes about gravitational attraction: "As for attraction, it is clear that this was employed by Newton, not as a true and physical quality, but only as a mathematical hypothesis" (DM 17).[17] Dynamic principles "serve mechanics and computation" (DM 18). "But," Berkeley continues, "it is one thing to serve computation and mathematical demonstrations, and another to exhibit the nature of things" (ibid.). Physical science is not to grasp the causes or essences of bodies but to mathematically model general theories that correspond to particular phenomena. As Downing (2005: 249) argues, to Berkeley "the theory as a whole serves as an instrument or calculating device for making kinematic predictions."

Furthermore, Berkeley separates the domains of metaphysics and theology from physical science. According to him, natural philosophy does not, and it should not seek "the efficient causes of things" (DM 35). Postulating an unobservable, supposedly explanatory, cause of force is not the aim of physics:

> Because these things are not sufficiently understood, some unjustly repudiate mathematical principles of physics, evidently on the pretext that they do not assign the true efficient causes of things. When in fact it is the concern of the physicist or mechanician to consider only the rules, not the efficient causes, of impulse or attraction, and, in a word, to set out the laws of motion: and from the established laws to assign the solution of a particular phenomenon, but not an efficient cause.
>
> DM 35

If these two interpretative steps—holding theories of physics as calculating devices and separating metaphysics and theology from physical science—are taken, the apparently occult feature of gravity vanishes. The term "force" is useful in organizing our experience concerning the motion of bodies, although we do not have a metaphysically valid idea of the thing "force" (Hight 2010: 18). Berkeley boldly renders Newtonian dynamics to instrumentalist direction (DM 36–7).[18] Even if the underlying cause of gravity is not known (which Newton and Berkeley can both be seen concurring with) and even if the law of universal gravitation, including its postulated concepts, lacks literal semantic correspondence to physical reality (which is Berkeley's position), it does not undermine the empirical success of Newton's physics.[19]

I think Hume also interprets physical forces—not causal relations— instrumentally. In footnote 16 to the first *Enquiry*, he notes that "when we talk of gravity, we mean certain effects, without comprehending that active power." This physical concept is an instrument: as if it provides a cause which refers to an effect: "the idea of power is relative as much as that of cause; and both have a reference to an effect" (EHU 7.29; SBN, fn. 17). To quote from Millican (2003: 145), "the ascription of powers to objects has considerable instrumental value," even though Hume suggests that the concept of force is "the unknown circumstance of an object." Forces are imperceptible, but they function as meaningful instruments in providing a phenomenal account for the laws of motion. The effect can be predicted by measuring the cause: "the effect is the measure of power" (ibid.). For Hume, the term force is a mathematical instrument which enables us to measure its effect, the change of motion of a body: "The degree and quantity" of an effect "is fixed and determined" by a force or power (ibid.). Besides this we do not comprehend what the causally efficacious parts are.

Here I wish to clarify that, in my interpretation, I do not claim that Hume would deny the existence of forces. Hume does not also eschew causal talk about forces, like Berkeley seems to do. And as the textual evidence provided suggests, Hume by all means thinks that causal relations are discovered. This is a realist position, of some sort. My contention is that Hume is agnostic about unobservable causes: he does not affirm or deny the existence of entities whose operations go beyond observed constant conjunctions. We are justified in accepting provable facts like the propositions concerning laws of nature in as much they are supported by past uniform experience, even though entities like force of gravity, energy, and the like to which these laws appeal are (in current science) imperceptible.[20] Future study of nature, with improving microscopic technology, or by any means which make microstructure observable, might disclose the causally efficacious parts of bodies. Beyond experience of constant conjunctions and the mathematical dimensions we do not comprehend what forces are. But Hume does not beforehand rule out the existence of forces. The physics of his time had mathematized forces, although the precise nature of these forces was still an open question. Hence Hume is not an anti-realist about forces and powers.

The Relation of Mathematics to Nature

To address Hume on the topic of the relation of mathematics to nature, this chapter is composed as follows. I first show the categorical difference between the two propositions divided by Hume's fork (henceforth HF). To do this, it is important to trace HF back to its foundation, to wit, the doctrine of relations. Then I turn to the problem of mixed mathematics. I first review the history of the notion of mixed mathematics briefly. HF provides an exhaustive and all-encompassing distinction among propositions of knowledge. This leads to the following difficulty: Is applied mathematics necessary and certain like pure mathematics, or is it contingent and probabilistic like factual knowledge? I argue that it is factual, because the application of mathematics in physics assumes the uniformity principle (UP), whereas pure mathematics does not. In conclusion, I advance the main claim of this chapter: the Humean view is that we should remain agnostic about whether reality has an underlying invariant structure that can be identified with mathematics. Hume's position on mathematics' relation to nature stands in contrast to Galileo's famous dictum, roughly that the book of nature is written in mathematical (more precisely geometric) symbols. Instead of positing a mathematical structure, Hume thinks that factual propositions benefit from mathematical formulation as they become affiliated with several epistemic virtues, like precision and predictability.

The Doctrine of Relations and Hume's Fork

Although HF in the first *Enquiry* is a statement of Hume's mature epistemology and philosophy of mathematics, the actual textual evidence for the distinction between mathematical and empirical propositions is scarce. Hume lays out his views on the category of a mathematical proposition in one paragraph. Then he contrasts relations of ideas to matters of fact, again using only one paragraph. The rest of the first part of section four elaborates the category of fact, touching upon mixed mathematics in the thirteenth paragraph. All in all, the argument for the separability of pure mathematics and factual knowledge (that incorporates applied math, as I will argue) in the first *Enquiry* is extremely short. Therefore, before delving into HF, it should be made clear what the foundation for this distinction is. As scholars (Cohen 1977; Millican 2017) have established, there is a connection between Hume's doctrine of relations as he presents it in the first Book of the *Treatise* (1.3.1), and HF as it appears in the *Enquiry*. A significant fact about HF is that it divides propositions with respect to relations. The two propositions concern either relations of ideas, or relations between species of objects. Yet in the first *Enquiry* Hume remains silent about what relations are. Hence the only way to understand HF properly is to inquire into his doctrine of relations.

Philosophical Relations

In the first Book of the *Treatise*, Hume notes that there are two types of relations, natural and philosophical. In distinguishing the two, he leans on the concepts of association and comparison. The term "relation" can be understood in two different senses:

> Either for that quality, by which two ideas are connected together in the imagination, and the one naturally introduces the other, after the manner above-explained; or for that particular circumstance, in which, even upon the arbitrary union of two ideas in the fancy, we may think proper to compare them. In common language the former is always the sense, in which we use the word, relation; and 'tis only in philosophy, that we extend it to mean any particular subject of comparison, without a connecting principle.
>
> T 1.1.5.1; SBN 13

In natural relations, the mind conceives some relations associatively. An example of such a relation might be the following case. I think that I am alone at home. I hear my friend's voice in the next room. Hearing the voice causes an auditory idea, which is associated with the thought of my friend; that is, the idea of my friend in the mind. Such association between my friend's voice and the image I have in my mind is not based on reflective comparison between ideas. Rather, the relation between the two is natural: I am naturally inclined to think that my friend is in my home. This is different from philosophical relations, in which the mind makes a judgment about relations; they are "subject of comparison" (T 1.1.5.1; SBN 13). As Hume's distinction between certain and probable relations concerns philosophical relations, and this particular divide is at the background of HF's categorical distinction, I will limit myself to studying philosophical relations only. Hume introduces seven philosophical relations (T 1.3.1; SBN 69–70). They can be divided into two groups. The first encompasses proportion in quantity or number (algebraic and arithmetic relations), degrees in quality, and contrariety. The second incorporates resemblance, identity, relations of space and time, and the relation of cause and effect.

The first criterion Hume mentions for the grouping is this: philosophical "relations may be divided into two classes; into such as depend entirely on the ideas, which we compare together, and such as may be chang'd without any change in the ideas" (T 1.3.1.1; SBN 69–70). Somewhat surprisingly, the example Hume gives for the first type of a relation is Euclid's sum-angle proof. In his original grouping, Hume mentions proportion in quantity and number, evidently meaning algebra and arithmetic. Nevertheless, his argument is that we may demonstrate that the angles of a Euclidian triangle equal two right angles, and this depends only on the relevant ideas of a triangle and of two right angles. The spatial or temporal interval of the two ideas does not matter. The sizes of the geometric figures do not matter for the relation of equality of their ideas. Also, the temporal interval is irrelevant: the first idea might be presented much earlier than the second idea, but this does not alter the relation itself in any way.

This is not the case with the second group of relations. Say that my cap and coffee mug are both laying on the table five inches apart. If I move the cap, and the mug stays where it is, the relation of space changes. Likewise, moving the cap is done in the course of time, so the temporal relation is also altered. Such variant relations do not depend solely on the comparison that the mind makes, but on the actual spatiotemporal locations of the compared objects (or events) in question.

After introducing this divide, Hume asserts that there are four relations which are dependent solely upon the comparison of ideas. Only these "can be the objects of knowledge and certainty" (T 1.3.1.2; SBN 70). The first three, resemblance, contrariety, and degrees in quality, are knowable by immediate perception; that is, intuition. I know immediately that a photo of my friend resembles my friend. I also know immediately that "It rains" is opposite to "It does not rain." When it comes to degrees in quality, Hume issues a word of caution. Slight differences in colors, senses of taste, and temperatures are difficult to observe, but considerable differences, say, a difference between a summer's day in San Fernando Valley and Christmas Eve in Lapland, are "easy to decide" by immediate perception, "without any enquiry or reasoning" (ibid.).

Sequences of intuitions constitute arithmetic and algebraic demonstrations. To explain this point, consider the following elementary propositions:

$1+1=2$

$2=2$

$a=a$

$2x=10$

$x=5$

$2\times5=10$

$10=10$

$a=a$

Hume thinks that algebraic and arithmetic calculations and proofs proceed by a comparison of ideas that form a unity. This is one clear and distinct idea in the mind. In these branches of mathematics,

> we are possest of a precise standard, by which we can judge of the equality and proportion of numbers; and according as they correspond or not to that standard, we determine their relations, without any possibility of error. When two numbers are so combin'd, as that the one has always an unite answering to every unite of the other, we pronounce them equal.
>
> T 1.3.1.5; SBN 71

In the last sentence of the above paragraph, Hume adds that as geometry does not work with the same standard of precision in which the mind can perceive equalities or inequalities among the compared discrete and exact ideas, it cannot yield the same kind of impeccable certainty as algebra and arithmetic do. Hume is however notoriously enigmatic on the status of geometry. In the *Treatise* 1.3.1.1, he first presents geometry as concerning the relations of ideas, but after introducing algebra and arithmetic in 1.3.1.5, he notes that geometry is not an exact science. This is analogous to his treatment of geometry in the first *Enquiry* as well. In 4.1 (SBN 25) he groups together geometry, algebra, and arithmetic, and notes that although "there never were a circle or triangle in nature, the truths, demonstrated by Euclid, would for ever retain their certainty and evidence." In this paragraph, he insinuates that geometry is impeccable. However, at the end of his work, he submits that it seems to him "that the only objects of the abstract sciences or of demonstration are quantity and number [...] As the component parts of quantity and number are entirely similar, their relations become intricate and involved" (EHU 12.27; SBN 163). Here Hume restricts the domain of the demonstrable to algebraic and arithmetic propositions, and repeats the criterion of exactness that he formulated in the *Treatise*. Geometry is not demonstrable because it does not lead to a unity between two calculable ideas. It is not clear what Hume's actual position about geometry is.[1] As I feel that this inconsistency is not of central concern to this book, I shall pass this point without further ado, and refer to mathematical propositions as relations of ideas, whether they are geometric, algebraic, or arithmetic.

In the second group of philosophical relations, Hume includes identity, spatial and temporal relations, and the relation of causation. These relations are not intuitive; they cannot be immediately perceived. The identity of an object does not remain the same over time. An object at time t_1 might be different at a later time t_2. Hume backs this up with his famous argument for diachronic personal identity.[2] In accordance with his copy principle, Hume's premise is that "I never can catch *myself* at any time without a perception, and never can observe any thing but the perception" (T 1.4.6.3; SBN 252). As perception is divided into two, the (complex) idea of self is derived from (particular) sensible impressions. But any impression that "gives rise to the idea of self"

> must continue invariably the same, thro' the whole course of our lives; since self is suppos'd to exist after that manner. But there is no impression constant and invariable. Pain and pleasure, grief and joy, passions and sensations succeed each other, and never all exist at the same time.
>
> T 1.4.6.2; SBN 251–2

Spatial and temporal relations are not intuitive, because they change when the locations of objects are altered. The last item on the list that Hume treats, the relation of causation, cannot be perceived, either. The reason is that causation is a relation; it is not something to be located in an individual object or event (this assumes the argument, established in the previous chapter, that we do not perceive causal powers). Causation is a relation pertaining to the objects. To continue with the example of the cap and the coffee mug, mere perception of the two does not tell us whether they are causally related or not. The tactile and visual impression causes the complex ideas of the cap and the mug. These are two distinct ideas. "The mere examination of two ideas present in our mind is not enough to tell whether or not they stand in the causal relation," notes David Owen (1999: 93). The relation of causation is not in the objects. We do not identify causation with perception, but with experience: "the power, by which one object produces another, is never discoverable merely from their idea, 'tis evident *cause* and *effect* are relations, of which we receive information from experience" (T 1.3.1.1; SBN 69–70).

The seven philosophical relations can be classified as follows:

Table 1 The seven philosophical relations

Invariant, provided that the compared ideas remain the same		Variant, although the compared ideas remain the same	
Certain		Probable	
Intuition	Resemblance	Reasoning	Identity
Intuition	Contrariety	Reasoning	Space and time
Intuition	Degrees in quality	Reasoning	Causation
Demonstration	Proportion in quantity and number		

For this chapter, the relevant items in the classification are mathematical and causal relations. Already in the *Treatise*, Hume thinks that mathematics is subject to demonstration, because the component ideas of theorems are intuitively equal. (In the first *Enquiry* (12.27; SBN 163), he writes that "as the component parts of quantity and number are entirely similar, their relations become intricate and involved.") Causal reasoning is the foundation of factual knowledge, which is subject to different degrees of probability. The dichotomy of mathematical demonstrability and causal probability in the *Treatise* forms a basis for HF in the first *Enquiry*.

Hume's Fork

A remarkable thing about Hume's fork (HF) is that it is a distinction among propositions. Already, in the first book of the *Treatise*, Hume argued for a dichotomy between certain and probable relations. As the doctrine of relations in the *Treatise* is a background for HF in the first *Enquiry*, propositions and relations must be tightly connected. In the second Book of the *Treatise*, Hume suggests that there are two types of relations that constitute propositions which can be true: "Truth is of two kinds, consisting either in the discovery of the proportions of ideas, consider'd as such, or in the conformity of our ideas of objects to their real existence" (T 2.3.10.2; SBN 449).[3] In the third Book of the *Treatise*, he makes a similar claim, insisting that "truth or falshood consists in an agreement or disagreement either to the real relations of ideas, or to real existence and matters of fact" (T 3.1.1.9; SBN 458). Finally, in the first *Enquiry*, Hume introduces the two propositions which concern either relations of ideas, or relations between species of objects or events. In his own words:

All the objects of human reason or enquiry may naturally be divided into two kinds, to wit, *Relations of Ideas*, and *Matters of Fact*.

EHU 4.1; SBN 25

All reasonings may be divided into two kinds, namely demonstrative reasoning, or that concerning relations of ideas, and moral reasoning, or that concerning matter of fact and existence.

EHU 4.18; SBN 35

Relations of Ideas

The first class of "reasoning," or object "of human reason or enquiry," consists of propositions concerning relations of ideas. As established in T 1.3.1 (SBN 69–73), these are either intuitively certain, or demonstrable by a sequence of intuitions. At the end of the first *Enquiry* (12.27 SBN 163), Hume adds that not only mathematical but also definitional truths founded on conventions can attain demonstrative certainty: "to convince us of this proposition, *that where there is no property, there can be no injustice*, it is only necessary to define the terms, and explain injustice to be a violation of property." Hume's rhetoric on syllogistic logic is somewhat cryptic. I think David Owen (1999: 107) is correct to point out that in the *Treatise* he drops "talk of syllogisms" altogether. This might not be the case in the first *Enquiry* (12.27; SBN 163). However, Hume accepts syllogisms only from the viewpoint of his theory of ideas; he is not championing any formally valid deductive modes of inference.[4] As this chapter pursues an adequate understanding of Hume's philosophy of mathematics, his conception of mixed mathematics, and the status of mathematics in science, I will limit my study to mathematical propositions only.

Mathematical propositions express nothing but the relations between ideas of figures, quantities, and numbers. As Hume writes: "the sciences of Geometry, Algebra, and Arithmetic; and in short, every affirmation, which is either intuitively or demonstratively certain [...] is a proposition, which expresses a relation between figures, quantities, or numbers" (EHU 4.1; SBN 25). The mind makes a judgment on whether the "component parts" of a mathematical proposition represent ideas that stand in an equal or unequal relation to one another (whether they become "involved" with each other or not, see EHU 12.27; SBN 163). Hume writes in the *Treatise* (1.2.4.21; SBN 46): "equality is a relation, it is not, strictly speaking, a property in the figures themselves, but arises merely from the comparison, which the mind makes betwixt them." The truth of a mathematical proposition is understood by comparing the relevant ideas of a given mathematical proposition. As this comparison is based on intuition, mathematics is demonstrable.

Hume argues that mathematical theorems are discoverable "by the mere operation of thought" (EHU 4.1; SBN 25). They are *a priori* truths. Millican (2007: xxxvi) explains Hume's position: "What makes a truth *a priori* is that it can be justified without appeal to experience, purely by thinking about the ideas involved." In judging the truth of mathematical propositions, we are relying solely on the mind's capability of comparing ideas with each other: "Thus as the necessity, which makes two times two equal to four, or three angles of a triangle equal to two right ones, lies only in the act of the understanding, by which we consider and compare these ideas" (T 1.3.14.23; SBN 166).

In the quote above, Hume maintains that mathematical truths are necessary. I think we can find two senses of necessity in his work: something is necessary if it is 1) true at all times, and if its 2) negation is inconceivable. Hume uses the notion of necessity in the sense of 1) in the first *Enquiry's* initial paragraph: "Though there never were a circle or triangle in nature, the truths, demonstrated by Euclid, would for ever retain their certainty and evidence." Necessary truths are eternal truths. This is already evident from the contrast Hume makes in the *Treatise's* doctrine of relations. Spatial and temporal relations are categorically different from relations that depend entirely upon ideas. Relations of ideas are independent of spatial or temporal dimensions. Euclid's sum-angle proof was true in Antiquity, it is true now, and it will be true in the fourth millennium. It is true in Greece, different continents, on planet Mars, and in galaxies far away from ours.

Hume also refers to necessity in the sense of 2), that is, mathematical theorems are what they are, and we cannot conceive them otherwise. In the first *Enquiry*, Hume thinks that the negations of true propositions of mathematics are inconceivable contradictions among ideas:[5] "Every proposition, which is not true, is there [in the proper science of mathematics] confused and unintelligible. That the cube root of 64 is equal to the half of 10, is a false proposition, and can never be distinctly conceived" (EHU 12.28; SBN 164 [My additions in the square brackets.]). There is still some confusion about the way Hume understands the negations of true mathematical propositions to be inconceivable. The confusion lies in the fact that in mathematical demonstration, that is, in the method and the process of proving a conjecture, one does not rely on showing the inconceivability of the negations of propositions. Consider, for example, a modernized version of Goldbach's Conjecture from the year 1742: Every even number greater than 2 can be expressed as the sum of two prime numbers (Millican 2017: 33). We can perhaps imagine that this conjecture is false. But Hume's point is that the inconceivability of the negations of propositions is a criterion for something to count as demonstrable. In the *Dialogues* (9.5; KS 189), Hume explains this in the line of Demea: "Nothing is demonstrable, unless the contrary implies a contradiction." Hume's position is that the negations of mathematical theorems (and demonstrable propositions in general) should not involve contradictions. He does not think that this plays any epistemic role in demonstration, that is, proof in the mathematical sense. In the formulation of Millican (2017: 36), "it might be argued" that

> we can only conceive in the weak sense what it would be for Goldbach's Conjecture to be true or false, since we can't conceive clearly and distinctly of its truth—quite obviously given its infinite nature—until we have found a proof, or of its falsehood until we have found a counterexample (i.e. an even number greater than 2 which cannot be expressed as the sum of two primes). Thus only after we have achieved the means of establishing its truth or falsehood will we be able to grasp such a mathematical conjecture sufficiently clearly to "see" its modal status.

Moreover, it is important to clarify what Hume means by contradiction. In his view, the term does not denote a logical contradiction, such as "It rains" and "It does not rain," which are mutually exclusive. In Hume, contradiction means a confusion that cannot be clearly and distinctly conceived by the mind: "'Tis in vain to search for a contradiction in any thing that is distinctly conceiv'd by the mind. Did it imply any contradiction, 'tis impossible it cou'd ever be conceiv'd" (T 1.2.4.11; SBN 43). The same point can be expressed in a propositional way, as Hume writes in the Abstract (18; SBN SBN 652f.) to the *Treatise*: "When a demonstration convinces me of any proposition, it not only makes me conceive the proposition, but also makes me sensible, that 'tis impossible to conceive any thing contrary. What is demonstratively false implies a contradiction; and what implies a contradiction cannot be conceived."

Although in this quote Hume models his position in terms of propositions, it should be emphasized that contradiction is fundamentally inconceivability among ideas. This is because propositions are made of ideas. In this sense, the negations of mathematical theorems are contradictory in Hume's theory. False mathematical utterances involve at least two confusing and incompatible ideas that do not form a unity, so the mind cannot conceive them clearly and distinctly.[6]

Matters of Fact

With respect to "all other enquiries of men," they "regard only matter of fact and existence" (EHU 12.28; SBN 163 f.). Matters of fact "are evidently incapable of demonstration" as "no negation of a fact can involve a contradiction" (EHU 12.28; SBN 163–4). Factual propositions are not dependent only on the ideas that the mind compares (Owen 1999: 83). By repeated experience, custom, habit, and natural instincts, two species of objects, such as flame and heat, and snow and cold, are related to each other. The relation between these types of objects is not discoverable by intuition, or demonstration, or by any type of *a priori* argumentation (EHU 4.7; SBN 28). We do not acquire factual information by merely comparing ideas between each other. Our knowledge concerning the relations between species of objects or events is founded on causality, which is founded on experience (Abstract 8; SBN 649, EHU 4.14; SBN 32).[7] Hume is explicit that the source of causal relations is experience: "'tis evident cause and effect are relations, of which we receive information from experience" (T 1.3.1.1; SBN 69). In the first *Enquiry*, we find a battery of arguments that corroborates this point:

> I shall venture to affirm, as a general proposition, which admits of no exception, that the knowledge of this relation is not, in any instance, attained by reasonings *à priori*; but arises entirely from experience.
>
> EHU 4.6; SBN 27

> I say then, that, even after we have experience of the operations of cause and effect, our conclusions from that experience are not founded on reasoning, or any process of the understanding.
>
> EHU 4.15; SBN 32

> It is only experience, which teaches us the nature and bounds of cause and effect, and enables us to infer the existence of one object from that of another.
>
> EHU 12.29; SBN 164

In reasoning regarding matters of fact, we proceed "upon the supposition, that the future will be conformable to the past" (EHU 4.19; SBN 36). But "the contrary of every matter of fact is still possible" (EHU 4.2; SBN 25). Hume argues that there is no contradiction in stating that the course of nature could radically change, that some familiar objects could be attended by certain unusual effects. It is distinctly conceivable that there would be snow and frost in July, and heat in January in Finland (EHU 4.18; SBN 35); it is distinctly conceivable that unsupported objects would not fall straight to the ground by the force of gravity (EHU 4.9; SBN 29); and it is indeed distinctly conceivable that a struck billiard ball would not continue its motion, to follow Newton's second law, "in the straight line in which that force is impressed" (EHU 4.10; SBN 29; Newton 1999: 416). But these are all questions of probability, and no matter of fact is subjected to the principle of contradiction (see EHU 12.28; SBN 168, and DNR 9.5; KS 189).

Interlude: Skepticism and the Uniformity Principle

Before proceeding to the problem of mixed mathematics, there is one problem that should be tackled. Consider the title of the section in which Hume introduces his fork: "Sceptical Doubts Concerning the Operations of the Understanding." He first speaks of propositions concerning "matters of fact and real existence." Such notions are clearly not skeptical. Rather, such talk indicates that we are able to reason about causal relations among real objects and events (not just figments of our imagination, for example), and such causal reasoning produces new information about the natural world. But, eventually, the second part of the chapter reaches a skeptical conclusion. Hume proposes a question to himself. What evidence is there for the uniformity principle (UP)? How do we know that the future conforms to the past, and that similar causes have similar effects? Hume gives "a negative answer to the question here proposed" (EHU 4.15; SBN 32).

As Millican (2007: xxxviii–xxxix) observes, Hume's skepticism concerning UP is understandable in tandem with his introduction of the three different kinds of evidence. The textual source is his "A Letter from a Gentleman to his Friend in Edinburgh" from the year 1745, written around the same time as the first *Enquiry*. In the letter, Hume maintains that "It is common for Philosophers to distinguish the Kinds of Evidence into intuitive, demonstrative, sensible [sensory], and moral [inductive]" (L 22). As he maintains such conception of the nature of evidence, UP is not justifiable for the following three reasons:

- *There is no intuitive or demonstrative evidence for the uniformity principle.* There is nothing intuitive about the similarity between the past and the future. We can conceive the one without conceiving the other. UP is not demonstrable.
- *There is no sensible [sensory] evidence for the uniformity principle.* We do not perceive a causal power, or an underlying mechanism, which necessitates that similar causes have similar effects and hence ground the uniformity of nature.
- *There is no moral [inductive] evidence for the uniformity principle.* We cannot experience the similarity between the past and the future. For instance, I do not have any experiences that yesterday is like tomorrow.

The skeptical Hume concludes that inferring the similarity of past and future

is not intuitive; neither is it demonstrative: Of what nature is it then? To say it is experimental, is begging the question. For all inferences from experience suppose, as their foundation, that the future will resemble the past, and that similar powers will be conjoined with similar sensible qualities [...] It is impossible, therefore, that any arguments from experience can prove this resemblance of the past to the future; since all these arguments are founded on the supposition of that resemblance.

                                                    EHU 4.21; SBN 36–8

There is no reason as to why we take a step from the past observed instances to the future ones. But we still *make* the step. Here what counts as "we" is much more inclusive than just "normal human beings" (say, healthy adults). Children and non-human animals also assume the uniformity of nature. Hume argues that they could not derive this from "any process of argument or ratiocination" (EHU 4.23; SBN 39). In the section "Of the Reason of Animals," he writes: "Animals, therefore, are not guided in these inferences by reasoning: Neither are children: Neither are the generality of mankind, in their ordinary actions and conclusions: Neither are philosophers themselves, who, in all the active parts of life, are, in the main, the same with the vulgar, and are governed by the same maxims" (EHU 9.5; SBN 106–7).

The uniformity assumption "is so essential to the subsistence of all human creatures, it is not probable, that it could be trusted to the fallacious deductions of our reason." Instead of slow deductive reasoning, nature has "implanted in us an instinct, which carries forward the thought in a correspondent course to that which she has established among external objects" (EHU 5.22; SBN 55).

The point Hume makes about "the fallacious deductions of our reason" motivates the skeptical, deductivist reading (for example, Stove 1973 and Popper 2002). Past and future, similar causes and similar effects, are logically distinct. There is no deductive justification for inductive inference. Such a skeptical reading maintains that inductive inference is invalid. Famously, Popper argued that neither in our everyday lives nor in scientific inquiry do we make inductive inferences. He took Hume to have shown that no inquiry is based on induction. Instead of Humean induction, Popper proposed that science proceeds with conjectures and refutations. It applies deductive *modus tollens* logic: hypothesis entails predictions, and critical testing falsifies hypothesis. If the hypothesis survives rigid testing, it will become a corroborated theory.[8]

However, there is a way to respond to such skeptical challenge without trying to justify the UP. As Millican (2003: 125–6) argues, one does not have to justify an inference from past to future, or causes to effects in a way that conforms to deductive logic. We may adopt a "far more modest principle [...] then the connexion between cause and effect must be at least to some extent non-arbitrary, and an examination of the cause must be able to yield some ground, however slight, for expecting that particular effect in preference to others" (Millican 2003: 126).

In Hume's words, "from causes, which appear *similar*, we expect similar effects. This is the sum of all our experimental conclusions" (EHU 4.20; SBN 36). This epistemically optimistic observation is consistent with an instrumentalist philosophy of science. We can understand mathematical dimensions of causes, and consequently predict future outcomes.

The Application of Mathematics

As all knowable propositions fall into two classes which concern two distinct types of relations, Hume insinuates that these types of propositions cannot be legitimately connected with each other. HF is an all-encompassing classification of propositions, and the distinction it implies is a dichotomy: exhaustive and mutually exclusive distinction among propositions of knowledge. Mathematics is confined to the realm of ideas: "It seems to me, that the only objects of the abstract sciences or of demonstration are quantity and number, and that all attempts to extend this more perfect species of knowledge beyond these bounds are mere sophistry and illusion" (EHU 12.27; SBN 164). Here Hume implies, among other things, that the necessity and certainty that are typical of mathematics cannot be extended to concern causal relations between physical objects. The proper objects of mathematical propositions are quantity, number, and figure. Judgments concerning relations between the component ideas of mathematical propositions are necessary and certain. Since extending "this more perfect species of knowledge beyond these bounds are mere sophistry and illusion," no causal relation, which is the founding relation in propositions concerning fact or existence (EHU 4.4; SBN 27, 4.14; SBN 32, 4.19; SBN 35), can be known to hold necessarily and certainly. Thus HF entails a dichotomous classification of propositions.

What about Hume's treatment of mixed mathematics in the first *Enquiry* (4.13; SBN 31)? It seems that reconciling mixed mathematics with HF poses a significant problem. Before rushing into Hume's treatment of mixed mathematics, it is useful to briefly look at the relevant history of this concept. According to Gary I. Brown's (1991: 81) study, the concept of mixed mathematics can be traced at least as far back as Francis Bacon's ([1605]1808) work *Of the Proficience and Advancement of Learnings*. In this work Bacon presented his tree of knowledge, the model of the branches of human learning. The model explicitly distinguishes pure and applied math. Sayaka Oki (2013: 83) shows that Bacon's view had predecessors, such as Adriaan van Roomen, Rudolf Snellius, and Petrus Ramus. Van Roomen explicitly used the notion of *mathematica mixta* in his 1602 work *Universae mathesis idea*. The intellectual background of the early modern conceptions of mixed mathematics was in the Aristotelian tradition of thinking about the difference between pure and applied mathematics. Douglas M. Jesseph (1993: 13) depicts the Aristotelian thinking as follows: "If the objects of mathematics are abstracted from the contents of the physical world, then we can take pure mathematics to be concerned with fully abstract objects and applied mathematics to treat partial abstractions which retain some of the sensible qualities of material objects."

The core idea of mixed mathematics is that it took its principles from pure mathematics and applied them to physical reality. As pure mathematics was understood to be absolutely certain, consequently mathematical demonstration about the physical world would also be absolutely certain. This kind of treatment of the application of mathematics was apparent already in Euclid's *Optics* and in Archimedes' *Equilibrium of Planes*. If demonstrations about optical ratios like the relationship between distance and angle, or dynamic relations like torque, were subject to a mathematical proof, then automatically such proof extends to the physical world. There was no problem about applying mathematics in the first place. If this kind of conception of mixed mathematics—in which there is no categorical difference between pure and applied math—were correct, HF cannot be right. Hume denies that facts, whether expressed in mathematical terms or not, could be necessary and absolutely certain.

Regarding Hume's philosophy, the crux of the problem can be explicated as follows. The propositions of mixed mathematics concern both relations between ideas *and* relations between species of objects. The former concerns the relation of intuition and the latter the relation of causation. Intuitive relations, given their relata, are necessary and certain, whereas causal relations are probable and fallible. Hence propositions of mixed mathematics seem to be both necessary/certain and probable/fallible. This cannot be; propositions of mixed mathematics have to either concern relations of ideas or be factual. To which class would the propositions of mixed mathematics then belong, according to Hume?

Hume mentions mixed mathematics explicitly only once in all of his works. His own example of mixed mathematics concerns the law of conservation of momentum. In his interpretation,

> it is a law of motion, discovered by experience, that the moment or force of any body in motion is in the compound ratio or proportion of its solid contents and

its velocity; and consequently, that a small force may remove the greatest obstacle or raise the greatest weight, if, by any contrivance or machinery, we can encrease the velocity of that force, so as to make it an overmatch for its antagonist.

EHU 4.13; SBN 32

Hume's encapsulation, "the moment or force of any body in motion is in the compound ratio or proportion of its solid contents and its velocity," suggests that he conflates

$$moment\; \alpha\; (solid\; contents) \times (velocity) \approx \vec{P} = m\vec{v},$$

and

$$force\; \alpha\; (solid\; contents) \times (velocity) \approx \vec{F} = m\vec{v}.$$

As Twardy (2014: 28) indicates, this confusion is probably derived from Colin Maclaurin's 1748 textbook *An Account of Sir Isaac Newton's Philosophical Discoveries*. Newton's own definition for momentum in the *Principia* is the following: "Quantity of motion is a measure of motion which arises from the velocity and the quantity of matter jointly" (Newton 1999: 404). However, Hume's confusion is not relevant to the problem of mixed mathematics, as I intend to analyze it in this chapter. The proposition which defines momentum, $\vec{P} = m\vec{v}$, "translation of momentum is a product of the mass and velocity of an object," expresses a relation between ideas. As momentum is a conserved quantity, the proposition is informative on how momentum is "transferred" between real objects, such as between billiard balls in a game of pool.

But how can this be? According to the first Book of the *Treatise* (1.3.1; SBN 69–73), the relations of intuition and demonstration are categorically different kinds of relations compared with the relation of causation. On the one hand, the definitional proposition $\vec{P} = m\vec{v}$ "expresses a relation between these quantities" (EHU 4.1; SBN 26). As such, it is an object of intuition, and demonstration. It can be algebraically manipulated, and the proposition does not refer to anything external. On the other hand, conservation of momentum belongs to "the laws of nature," describing "the operations of bodies," which, "without exception, are known only by experience" (EHU 4.9; SBN 30). Since translation of momentum in a system of bodies is observed to be contiguous between two objects, and there is a temporal sequence between the objects, the mathematical formulation of the law satisfies the conditions that Hume assigns to causal inference (see section XV of the *Treatise*, "Rules by which to judge of causes and effects," and the Abstract (9; SBN 649) of the *Treatise*).

There is another problem in Hume's mixed mathematics. HF divides propositions with respect to contradictions of their negations: the negations of relations of ideas are inconceivable contradictions, whereas the negations of matters of fact are distinctly conceivable. If the rules of algebra are followed, the proposition $\vec{P} = m\vec{v}$ cannot be rendered, without a contradiction, to propositions $m = \vec{P}\vec{v}$, or $\vec{v} = \vec{P}m$. In Hume's theory, the negation of $\vec{P} = m\vec{v}$ would be an inconceivable contradiction among the component ideas of this proposition. But it is possible to conceive a situation when a cue ball hits the object ball in the game of pool, the object ball does not continue its motion to the direction of $\vec{P}$ but stays in position or moves in a direction other than $\vec{P}$. This indicates that referring to the principle of contradiction does not suffice to settle the issue of Hume's mixed mathematics. Rather, as I shall argue next, this requires UP.

The Dependency of Mixed Mathematics on the Uniformity Principle

In my interpretation, the reason why Hume does not allow demonstration to be extended to natural events is this: one has to presuppose the uniformity of nature. Instinctively and habitually we, both humans and non-human animals, assume that the future resembles the past. We infer "that the same events will always follow from the same causes" (EHU 9.2; SBN 105). The relevance of UP to Hume's conception of mixed mathematics is also echoed in his own statement: "Every part of mixed mathematics proceeds upon the supposition, that certain laws are established by nature in her operations" (EHU 4.13; SBN 31).

Matter of fact propositions, although expressed as mathematically formulated laws, are not based on reason and are justifiable neither by intuitive nor by demonstrative reasoning. We do not know their truth by a mere comparison of the relations of ideas, just by consulting our intellectual faculties. Factual reasoning requires the comparison of how objects are related in the actual world by a customary transition "from the appearance of a cause [...] to the effect" (EHU 7.29; SBN 76). And our knowledge concerning these causal relations, as Hume frequently argues in the first *Enquiry* (4.6; SBN 27, 4.15; SBN 32, 12.29; SBN 164), is not founded on, nor can be justified by, *a priori* reasoning. Consequently, even when matter of fact propositions are formulated mathematically, they are *a posteriori*.

One startling objection could still be made to this interpretation. Using a mathematically formulated proposition such as $\vec{P} = m\vec{v} = conserved$ enables one to deduce a contingent conclusion from contingent premises. To illustrate this, consider the following reformulation of an isolated system in which momentum is conserved. The system contains two objects, bodies A and B. An object A, which is in motion, collides with object B, which is at rest. The contingent premise consists of the initial conditions i, the momenta of $\vec{P}_A$ and $\vec{P}_B$. Alike, the final condition f, is a contingent conclusion. The initial and final matters of fact are all contingent, since the salient variables, the masses m_A and m_B, and the velocities $(\vec{v}_i)_A, (\vec{v}_i)_B, (\vec{v}_f)_A, and (\vec{v}_f)_B$, are all contingent. However, deducing the conclusion from the premise is not a contingent procedure: $\vec{P} = m\vec{v} = conserved$ can be, step by step, algebraically manipulated to determine the desired variable. As a mathematical proposition, conservation of momentum does not refer to anything external. It depends solely on the quantities it is composed of. The situation is like in Table 2 below.

With the inference above, it is possible to demonstrate, before observing the collision of A and B, what the motion of A and B will be. Although the premise of such an argument is contingent, and not in any way necessary, the deduction to the conclusion is necessitated by the given quantities, and hence the process is certain. It can be argued, then, that the demonstration in this case brings forward new information about some factual matter. Before any experience, that is, before perceiving what happens in the collision of objects A and B, we are able to demonstratively infer what happens to the system AB when A and B are conjoined. But how can this be, when Hume adamantly denies that "there is no demonstration [...] for any conjunction of cause and effect" (Abstract 11; SBN 650) that "enquires" regarding "matters of fact and existence [...] are evidently incapable of demonstration" (EHU 12.28; SBN 163), and that "there is an evident absurdity in pretending to demonstrate a matter of fact, or to prove it by any arguments *a priori*" (DNR 9.5; KS 189)?

To reiterate the main point of this section, the solution to the former mystery is this: *Given* UP, "the course of nature continues always uniformly the same," the premises transfer the truth to the conclusion. As De Pierris (2006: 305) argues, this inference is "licensed by the principle of the uniformity of nature." But in Hume's view, there is no demonstrative guarantee that "the future resembles the

Table 2 Demonstrative inference from a contingent premise to a contingent conclusion

Contingent premise	Demonstrative inference	Contingent conclusion
Momenta of A and B	$\vec{P}_i = \vec{P}_f$	Momenta of A and B
	Deduction of the desired variable, like $(\vec{v}_f)_A$ or $(\vec{v}_f)_B$.	

past" (See T 1.3.6.4–5; SBN 89, 1.3.12.9; SBN 134). We can conceive that nature will change its course, which is enough to rule UP out as an impeccable principle:

> Our foregoing method of reasoning will easily convince us, that there can be no *demonstrative* arguments to prove, *that those instances, of which we have had no experience, resemble those, of which we have had experience.* We can at least conceive a change in the course of nature; which sufficiently proves, that such a change is not absolutely impossible. To form a clear idea of any thing, is an undeniable argument for its possibility, and is alone a refutation of any pretended demonstration against it.
>
> T 1.3.6.5; SBN 89

As UP itself is not provable by intuition or demonstration, it must be that the propositions of mixed mathematics cannot be provable by intuition or demonstration, alone. This also explains a comment Hume makes in the first section to the second *Enquiry*, where he points out that theories about the laws of nature might be refuted, unlike pure, non-applied mathematical theorems: "Propositions in geometry may be proved, systems in physics may be controverted" (EMP 1.5; SBN 171). Propositions concerning laws of nature, although they do express relations between numbers, quantities, and figures, do not share the same necessity and certainty as propositions of pure arithmetic, algebra, and geometry. The former are dependent on UP; the latter are not.

The reason why Hume gives a high epistemic status to the laws of nature is that they signify a set of causes and effects which have "hitherto admitted of no exception" (EHU 6.4; SBN 58). In fact, applying mathematics does not guarantee certainty—it is not really mathematics that renders the laws of nature as high-class matters of fact. Rather, it is their regular, unexceptional occurrence. As Hans Reichenbach (1951: 159) clarifies Hume's position: "laws of nature are for him statements of an exceptionless repetition—not more."[9] Their epistemic status is similar, compared with other "common" causal facts of nature, such as our knowledge of fire having the attribute of burning or water having the attribute of drowning nonaquatic beings (see EHU 6.4; SBN 57).

The example of burning fire and drowning water illustrates how Hume understands the epistemic status of propositions of mixed mathematics. In the sixth section to the first *Enquiry* (6.4; SBN 57), Hume groups together both mathematically expressible matter of fact propositions (roughly, Newton's second law of motion, and the law of universal gravitation) and qualitatively expressible matter of fact propositions (fire burns, and water causes drowning to non-aquatic beings):

There are some causes, which are entirely uniform and constant in producing a particular effect; and no instance has ever yet been found of any failure or irregularity in their operation. Fire has always burned, and water suffocated every human creature: The production of motion by impulse and gravity is an universal law, which has hitherto admitted of no exception.

Later, in the tenth section of the first *Enquiry* (10.4; SBN 110), Hume indicates that the regular, unexceptional occurrence of these kinds of causal relations renders them "proofs."[10] His position can be sketched as in Table 3 below.[11]

Both propositions, quantitative and qualitative, count as "proofs" because they are instances of exceptionless repetitions. To Hume, "proofs" are high-class, non-necessary propositions about constant conjunction between two species of objects. It is not relevant, according to this classification, whether a matter of fact proposition is expressed quantitatively or qualitatively, since, as Hume points out in the fourth section of the first *Enquiry*: "all our reasonings concerning fact are of the same nature" (EHU 4.4; SBN 26, see also Abstract of the *Treatise* 10, SBN 650). Mathematical or not, the logic of inductive arguments is the same in both cases: matter of fact propositions presuppose UP as a latent premise.

It should be noted that there is still a difference in causal probabilities and proofs in Hume's account. He discusses this difference both in the *Treatise* (1.3.11.2; SBN 124) and in the first *Enquiry* (6, fn. 10). He asserts that proofs "exceed probability," being "entirely free from doubt and uncertainty." They "leave no room for doubt or opposition." Hume thinks that there are empirical proofs that human reason does not doubt, such as "all humans will eventually die," and "the sun will rise tomorrow." But these proofs are not absolutely certain, or necessary. The evidence of proofs is higher than probabilities, but it is not as high as in demonstrations. "Proofs" are thus both free of doubt and fallible.

Moreover, when Hume (T 1.3.11.2; SBN 124, EHU 6, fn. 10) introduces the tripartite categorization, which includes demonstration, proof, and probability, it would be false to think that this categorization is epistemically fundamental. As I have shown in this chapter, the fundamental categorization is the dichotomous

Table 3 Hume's classification of "proofs"

Quantitative	Qualitative
$\vec{F} \propto \Delta \vec{P}$	Fire burns.
$\vec{F}_G \propto \dfrac{m_1 m_2}{r^2}$	Water causes drowning to non-aquatic beings.

distinction of HF in the first *Enquiry*, which is grounded in the doctrine of relations in the first Book to the *Treatise* (1.3.1). HF is a distinction of kind, not degree. Although Hume understand proofs and probabilities to clearly differ in the degrees of their evidence (T 1.3.11.2; SBN 124), the proof/probability distinction is not a dichotomous distinction like HF. Hume's fundamental epistemic categorization is between relations of ideas, which are founded on the relation of intuition, and are thus capable of being demonstrated without any appeal to fallible experience, and matters of fact, which are founded on the relation of causation, and which do require a reference to fallible experience in their justification.

Mixed Mathematics Instantiates Epistemic Virtues

Hume classifies causal relations that can be expressed in mathematical terms in the same way as causal relations that are expressed in qualitative terms; he labels them as "proofs." As Deborah Boyle (2012: 158) points out, it is quite universally accepted in the secondary literature that there is no eventual rational justification for a belief in any causal inference. Even Hume's "proofs" require UP as a latent premise. This principle itself is a customary, habitual, and instinctive principle, not a principle founded on reason. However, many Hume scholars have emphasized the normative and constructive character of certain causal inferences (for example, Millican 1998; De Pierris 2001; Boyle 2012; Schafer 2014). Hume talks about "wisdom" and "good" sense, and insists that "a wise man [...] proportions his belief to the evidence" (EHU 10.4; SBN 110). For instance, "proofs" have a specific normative character not shared by mere "probabilities." The former are supported by the whole of past uniform experience, whereas the latter are not. Hume is committed to a normative claim: uniform past experience and repetition are virtues which should be appraised by a wise person.

Since mixed mathematics and qualitative "proofs" are on a par with respect to their certainty, the only difference between these provable causal propositions is that mixed mathematics can be associated with some epistemic virtues. Hume's rhetoric in T 2.3.3.2 (SBN 143) and EHU 4.13 (SBN 31) clearly esteems the application of mathematics. Hume allows that "mixing" mathematics enhances precision, or "accuracy of reasoning," and that it "assists experience" in making the discovery and application of laws of nature possible. Mathematics is very useful in mechanical operations: "Mathematics, indeed, are useful in all mechanical operations, and arithmetic in almost every art and profession"

(T 2.3.3.2; SBN 413f.). Hume thinks that it is simply a good thing that mathematics can be used and applied to a variety of different disciplines, such as physics, agriculture, building, and commerce. Hume's treatment of mixed mathematics, such as in his own example about conservation of momentum, allows that it is possible to make accurate predictions of motions of objects by *a priori* mathematical demonstration. But Hume's logic of induction indicates that predictability is still founded on past experience. The past experience, although it can be brought under a quantitative law, enables one to infer the yet unobserved future, in the exact same way as the past experience enables me to infer that when I put my finger in a flame I will feel pain and heat.

Hume denies that the necessity and certainty related to abstract mathematical reasoning could be extended to concern factual reasoning (EHU 12.27; SBN 164). Matters of fact are founded on causation, which is founded on experience. But Hume does allow that abstract mathematical reasoning can assist experience in the discovery and application of laws of nature (EHU 4.13; SBN 31). Thus the appropriate way to understand the epistemic status of propositions of Hume's mixed mathematics is that they neither instantiate necessity, nor do they increase certainty. They are matters of fact that represent epistemic virtues of precision, predictability, and usefulness.

Is the Book of Nature Written in the Language of Mathematics?

Finally, we can assess Hume on the relationship between mathematics and nature. In what sense does mathematics grasp the structure of nature? Galileo's (1965) famous paragraph in *The Assayer* implies that nature is mathematical or, more precisely, geometrical:

> Philosophy is written in this grand book—I mean the universe—which stands continuously open to our gaze, but it cannot be understood unless one first learns to comprehend the language and interpret the characters in which it is written. It is written in the language of mathematics, and its characters are triangles, circles, and other geometrical figures, without which it is humanly impossible to understand a single word of it; without these one is wandering about in a dark labyrinth.

The above paragraph is one of the most often-quoted in the history of science. Mario De Caro (1992: 1) notes that "the core of Galileo's thought is the mathematical conception (to be precise, the geometrical conception) of reality;

of the method of science; and of the forms of knowledge." As this chapter aims to interpret Hume on the relation between mathematics and nature, it suffices to briefly comment on Galileo's understanding that reality is mathematical.[12] Galileo assumes that space is geometrical. Bodies' non-geometric properties, like color, are subjective and not parts of objective reality. "Galilean ontology," De Caro (ibid.) argues, "is rigorously restricted to mathematizable, or better geometrizable, entities and properties." This geometric structure is the mathematical structure of reality.

Such a view about the mathematical nature of reality is, I will argue below, neither Hume's position nor contrary to it. He does not deny the mathematical nature of reality but rather maintains agnosticism about it. In Hume's account, mathematics is a useful tool which enables one to express the magnitudes of causes (such as forces) and effects (accelerations produced by forces) in a precise and predictable manner. In this broadly instrumentalist picture of the application of mathematics in science, Hume remains agnostic on whether there are invariant structures that go beyond perceptions of objects or events.

In the Galilean view, the geometric ratios of bodies make up the objective reality. This position presumes a distinction between primary and secondary qualities: extension is primary, whereas color, texture, temperature, and the like are secondary. Such a realist view maintains that the primary qualities are invariant and independent of human perception. Hume depicts the primary/secondary distinction in the last section of his first *Enquiry* (12.15; SBN 154): "It is universally allowed by modern enquirers, that all the sensible qualities of objects, such as hard, soft, hot, cold, white, black, *&c.* are merely secondary, and exist not in the objects themselves, but are perceptions of the mind, without any external archetype or model, which they represent." Then Hume goes on to suggest that primary qualities are only presumably "qualities of extension," eventually challenging the whole distinction:

> The idea of extension is entirely acquired from the senses of sight and feeling; and if all the qualities, perceived by the senses, be in the mind, not in the object, the same conclusion must reach the idea of extension, which is wholly dependent on the sensible ideas or the ideas of secondary qualities. Nothing can save us from this conclusion, but the asserting, that the ideas of those primary qualities are attained by *Abstraction*; an opinion, which, if we examine it accurately, we shall find to be unintelligible, and even absurd. An extension, that is neither tangible nor visible, cannot possibly be conceived.
>
> ibid.

Bereave matter of all its intelligible qualities, both primary and secondary, you in a manner annihilate it, and leave only a certain unknown, inexplicable *something*, as the cause of our perceptions; a notion so imperfect, that no sceptic will think it worth while to contend against it.

EHU 12.16; SBN 155

Given Hume's denial of primary/secondary distinction (or his Berkeleyan inspired criticism of Locke's representative realism), geometric proportions of bodies are not—or we do not have a way to ascertain—the fundamental structure of reality. Such an argument would violate Hume's copy principle. We get our (simple) ideas from sensory impressions; we do not have ideas about putatively mind-independent geometric ratios. Nature might be mathematical in its structure. Given that our thoughts are confined within our impression-based ideas, we cannot know this. We cannot even think things themselves, like mathematical structure, without sensory qualities.

Space and Time

Hume's philosophy of space and time is not, in the first place, related to physics. He is interested in the ideas of space and time. This is in accordance with the objectives of his science of human nature: to explain the origin and nature of the ideas we employ.

Granted, Hume's approach to space and time concentrates on human perception, rather than physical quantities. But this does not mean that Hume's views are independent of the natural philosophical tradition. To show this, I will proceed as follows. I first present Newton's absolutist argument and his criticism of Descartes' position on space. Then I move on to Hume's doctrines. I argue that Hume is closer to Descartes on the topics of space and the vacuum, and that he also discredits Newton's absolute universal time.

The Absolutist Argument

Newton makes his argument for absolute space and time in the Scholium to the first Book of the *Principia*, between the sections "Definitions" and "Axioms, or Laws of Nature." He begins by noting that our everyday vocabulary includes words like "space," "time," "place," and "motion." The problem is that "these quantities are popularly conceived solely with reference to the objects of sense perception." Such conception forms our presumptions about space, time, place, and motion. The popular conception of time draws from the Aristotelian account, according to which time is dependent on the motions of bodies and in some way on the existence of physical reality (Schliesser 2013: 89). Newton thinks this is a wrong starting point, and therefore he distinguishes the quantities into "absolute and relative, true and apparent, mathematical and common."

According to Newton, space is distinct from bodies, and time flows equably, independently of whatever happens to bodies. Space and time are not substances but absolute structures.[1] Absolute space is a three-dimensional, homogenous

Euclidian structure that extends to infinity in all its three directions. It is immaterial and hence independent of matter. In Newton's formulation: "Absolute space, of its own nature without reference to anything external, always remains homogeneous and immovable." In contrast to absolute space, "relative space is any movable measure or dimension of this absolute space." Absolute space is imperceptible, but relative space is observable and measurable. Newton defines place as "the part of space that a body occupies." Absolute place is a body's relation to absolute space, and relative place the body's spatial relation to other bodies (*Principia*, first Book, Scholium, 2–3).

Newton characterizes absolute time as follows: "Absolute, true, and mathematical time, in and of itself and of its own nature, without reference to anything external, flows uniformly and by another name is called duration." Time is absolute as its existence does not depend on anything else but itself, excluding God's existence. Absolute time is distinct both from bodies and from the absolute space. It flows equably, so the motions of bodies do not affect its pace in any way: "All motions can be accelerated and retarded, but the flow of absolute time cannot be changed," writes Newton.

Time has an absolute structure because it "flows uniformly." We do not observe the flow of time, and we cannot measure it with any timekeepers. But we can mathematically grasp the difference between equal and unequal temporal intervals. Thus, Newton's argument for absolute time relates to his laws of motion (DiSalle 2006: 20–1; Schliesser 2013: 89). This can be illustrated with motion diagrams in which two bodies are contained in and move within absolute space. The dots picture moments—a snapshot of the motion of a body—that have equal pairwise spatial relation. The distances between the dots represent temporal separations:

In a), a body is not subjected to a net force. It moves equal distances in equal times. The temporal intervals 1-2-3-4-5-6 are absolutely equal. In b), there is a constant non-zero net force applied to a body, so it moves at absolutely uneven temporal intervals. The temporal interval 1-2 is absolutely different from 2-3, which is different from 3-4, and which is respectively different from 4-5, and which in turn is different from 5-6. The diagrams are idealized, and they do not show exact proportions. However, this point is embedded in Newton's argument:

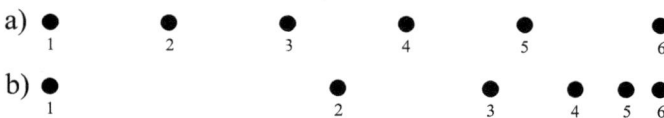

Figure 3 Equal and unequal temporal intervals.

we cannot perceive or measure absolute time, but we can mathematically grasp it. Hence Newton groups together absolute, true, and mathematical time, as opposed to "relative, apparent, and common time." The latter is "any sensible and external measure (precise or imprecise) of duration by means of motion; such a measure—for example, an hour, a day, a month, a year—is commonly used instead of true time" (*Principia*, first Book, Scholium, Definitions).

A familiar example of measuring and observing relative time is the ticking of a wristwatch. Our watches do not tick perfectly evenly. Absolute time makes sense to the notion of a more accurate/less accurate clock: the more evenly the clock ticks, the more accurate it is. Although "it sounds," as Tim Maudlin (2012: 15–6) has it,

> as if Newton is postulating some weird, ghostly, unfamiliar entities, but most people conceive of the physical world in terms of absolute space and time. For example, craftsmen and scientists continually try to improve the design of timepieces, to produce clocks that are ever more accurate and precise. But what is it for a clock to be "accurate"? What we want is for the successive ticks of the clock to occur *at equal intervals of time*, or for the second hand of a watch to sweep out its circle *at a constant rate*. But "equal" or "constant" with respect to what? With respect to the passage of time itself, that is, with respect to absolute time. Our natural, intuitive view is that a certain amount of absolute time elapses between the successive ticks of a clock, and the better and more accurate the clock is, the more similar these intervals are to one another. Swiss watchmakers, and designers of atomic clocks, are trying to get their devices to accurately measure something, and that something is not any sort of relative, observable time.

As Ducheyne (2001: 78) argues, Newton does not treat absolute time as mere idealization of clock time. Newton equates absolute and true time; the two are not necessarily the same.[2] Absolute time denotes the substantial existence of time. According to Ducheyne (2001: 86), Newton's reference to "true" time debunks Lawrence Sklar's (1990) representationalist view. Absolute time is not a mere imaginary postulate that is constructed to account for laws of motion, but a real mind- and body-independent structure. It has a role to play in Newton's broader onto-theological network (ibid.: 79).

Newton's philosophy of time is commonsensical when compared with an alternative view, like the "standard clock" definition of accuracy. According to this definition, a clock is accurate if it is in synchrony with the standard clock (Dowden 2018, Section 23). This approach leads to a very counterintuitive situation. In the light of relativistic physics, we might say that there is no true time, or time itself, as Newton thought. No clock is "truer" than any other clock, in the same way as no frame of reference is "truer" than any other. There is no one universal clock that ticks evenly, and with which we could compare our own local clocks. The choice of standard clock is conventional. It might be the atomic clock in Boulder, Colorado, the president's heartbeat, or the dripping of water from my in-laws' cottage roof. If the standard clock, whatever it might be, ticks once between the Big Bang and the 2019 New Year, and a million times between the 2019 New Year and our present time (assuming the commonsensical presentism!), the intuitive reaction is that there is something wrong with such a clock: it does not imitate true time but represents just our apparent measures of time.[3] The Newtonian view is in accordance with common sense, because a reference to absolute time explains how a temporal metric is grounded in time itself.

Newton's position on time is also very intuitive in another sense. Time is universal. Thus Newton puts the point in "De Gravitatione" (2004: 26) and the General Scholium of the *Principia*: "For we do not ascribe various durations to the different parts of space, but say that all endure simultaneously. The moment of duration is the same at Rome and at London, on the earth and on the stars, and throughout all the heavens [. . .] each and every indivisible moment of duration is *everywhere*."

So, if something happens "now," it happens "now" universally.[4] The order in which things happen, the relations of simultaneity and succession, are absolute, observer-independent. As noted earlier, absolute time flows equably, so it has a definite structure: all simultaneous events happen absolutely at the same time, because the time difference between them is zero, and all successive events happen at absolutely different times, as the time difference between them is a non-zero quantity (Earman 1989: 8–9). This renders Newton's conception again perfectly commonsensical: simultaneity is assumed as being an absolute relation between two events.[5]

Newton's absolute space and time are implicit in his laws of motion. In the original formulation of the second law in the *Principia* (Axioms, or Laws of Motion), Newton refers to "impressed forces." This is evidently a causal notion. Steffen Ducheyne (2012: Section 1.5) shows that there are relevant similarities between Newton's causal conception of scientific reasoning and the Aristotelian textbook tradition (as exemplified, for example, by Jacopo Zabarella's method of *regressus*[6]). A central idea in the Aristotelian tradition is that nature is causally ordered. In Newton's account, forces are causes, and changes of motion are effects. Causes are ontologically prior but epistemically secondary, whereas effects are ontologically secondary but epistemically primary.

Here, a brief example of such causal-scientific reasoning should suffice, as I have already treated an example like this in Chapter 1. I throw a rock, and I see its parabolic trajectory. The effect, that is the non-rectilinear path of the rock, is my epistemic access to the cause of the change of motion. Without the existence of the force of gravity, that is the ontologically primary feature in the causal process, the rock would fly straight. We do not see unobservable causes, the forces, but we infer them from their effects, the accelerations. In the Queries of the *Opticks*, Newton dubs this "the main Business of natural Philosophy," which is "to argue from Phaenomena without feigning Hypotheses, and to deduce Causes from Effects." The deduction of causes from effects is the analytic part of the inquiry. It is followed by the synthetic part, as Newton explains in Query 31:

> By this way of Analysis we may proceed from Compounds to Ingredients, and from Motions to the Forces producing them; and in general, from Effects to their Causes, and from particular Causes to more general ones, till the Argument end in the most general. This is the Method of Synthesis: And the method of Synthesis consists in assuming the Causes discover'd and establich'd as Principles, and by them explaining the Phænomena proceeding from them, and proving the Explanations.

In the synthetic part, we can explain the phenomena of motion—accelerations like projectiles, the tides, lunar and planetary orbits—by referring to the common cause, the force of gravity. The law of universal gravitation is the principle which explains changes of motion in the scale of our solar system, as Newton concludes the General Scholium: "And it is enough that gravity really exists and acts according to the laws that we have set forth and is sufficient to explain all the motions of the heavenly bodies and of our sea." Gravity also has a proximate cause, which Newton does not know, but the ultimate, most remote cause is a theistic God, which is apparent in the design argument of the General Scholium: "This most elegant system of the sun, planets, and comets could not have arisen without the design and dominion of an intelligent and powerful being." Gravity explains a good deal of motions in our solar system, but not the original order of planets and who set them in motion in the first place.

How is the causal-scientific reasoning related to space and time? An important aspect in Newton's absolutist argument is to show that his laws of motion are objective descriptions of material reality. If a body is impressed by a force, it absolutely, not just relatively, moves. To expound on this difference, we may consider the two following cases, kinematical and dynamical.

- *Kinematical case.* An observer stands at a platform, and another observer sits on a train moving with constant velocity. From the viewpoint of the first observer, she is standing still, and the train is moving. For her part, the second observer thinks she is at rest and it is the other who is moving.
- *Dynamical case.* I move a coffee mug on the table. The mug is at rest and an impressed force makes it move. I catch a baseball flying toward me. The ball is moving, and the application of a force stops it.

In the kinematical case, there is no change of motion. In the dynamical case, there is. If there is a net force impressed on an object, it truly, not just relatively, changes its state of motion. As Newton has it in the Scholium to the Definitions of his *Principia*:

> The causes which distinguish true motions from relative motions are the forces impressed upon bodies to generate motion. True motion is neither generated nor changed except by forces impressed upon the moving body itself, but relative motion can be generated and changed without the impression of forces upon this body.

In the cases of me shoving the coffee mug or stopping a baseball, the bodies in question truly move or come to a halt. These motions are not perspectival, unlike in the kinematical case.[7] In our vocabulary, these motions are not dependent on the selection of the inertial frame of reference. Newton does not explicitly use the concept of an inertial frame; instead, he relies on the concepts of force, laws of motion, and absolute space and time. For him the world has the frame of its own, which is the absolute Euclidian space.

To back up this point, Newton presents an argument from effects, not just from causes, with his rotating water bucket experiment and the thought experiment with revolving globes.

Bucket Experiment and Revolving Globes

Newton notes that "the effects distinguishing absolute motion from relative motion are the forces of receding from the axis of circular motion." He reports an experiment, which may be easily repeated. Fill a bucket half-way with water. Tie the handle of the bucket with a rope to the ceiling. Twist the bucket as much as the rope allows, and release it. We first observe that the surface of the water is flat. After a while, the friction of the rope affects the water, and the surface becomes concave. In this stage, the water and the bucket are both spinning, so they are at rest with respect to each other. As we see an upward motion, there is a force impressed onto the water, and the water is truly moving: "The rise of the water reveals its endeavor to recede from the axis of motion, and from such an endeavor one can find out and measure the true and absolute circular motion of the water, which here is the direct opposite of its relative motion." But we cannot detect this motion by comparing it with its immediate surroundings. Andrew Janiak (2013: 106) comments on Newton's inference. "Although we cannot detect absolute space itself, since Newton conceives of it as empty and imperceptible, we can detect the inertial effects of rotating bodies (in this case, the concavity of the water)."

The bucket experiment establishes a need to analyze true motion in terms of a reference to absolute space. For its part, the revolving globes example illustrates that although absolute space is not apparent to our senses, we may nevertheless infer the absolute quantity of motion from a dynamic system (Rynasiewicz 2014: Section 4). In Newton's example, there are two globes connected by a cord in an otherwise empty space. This example tries to establish the absoluteness of rotational motion with the following reasoning. If the globes are revolving around a common center of mass, they "try" (Newton uses the word "endeavor") to recede from the axis of motion. This is apparent in the tension of the cord. Whether the motion of the globes is clockwise or counterclockwise can be detected by accelerating or decelerating the globes, which results in either increment or diminution in the tension of the cord. The quantity of this tension, and consequently the quantity of circular motion "could be found in any immense vacuum, where nothing external and sensible existed with which the balls could be compared," Newton claims. Then he continues and makes a comparison between the globes and stationary objects, like remote fixed stars. Observing the motion of the two-globe system does not itself yet inform us whether the globes or the reference stars are truly moving. "But," Newton goes on to point out,

> if the cord was examined and its tension was discovered to be the very one which the motion of the balls required, it would be valid to conclude that the motion belonged to the balls and that the bodies were at rest, and then, finally, from the change of position of the balls among the bodies, to determine the direction of this motion.

In Newton's example, there is an observable tension in the cord. This produces a centripetal force, which indicates the absolute rotation of the globes. This absolute motion takes place even though the globes and the cord do not change their locations with respect to each other (Maudlin 2012: 23).

Newton's Criticism of Descartes

As is well known, in Descartes' metaphysics there are two substances: minds and bodies.[8] Only the latter substances are the subject matter of physics. Descartes shuns the Aristotelian-Scholastic explanations of natural phenomena, which employ (putatively) causally efficacious substantial forms. Edward Slowik (2017: Section 2) notes that Descartes wishes to replace "metaphysically suspect properties," like heat, weight, texture (in the sense of how does a body feel), with "quantifiable attributes of size, shape, and motion." Slowik (ibid.) continues: "Descartes intends to replace the 'mentally' influenced depiction of physical qualities in Scholastic natural philosophy with a theory that requires only the properties of extension to describe the manifest order of the natural world." Moreover, Descartes maintains that metaphysics is primary and physics secondary. This is evident in his widely known classification of the sciences (in early modern parlance, "philosophy") in the French "Author's Letter" to his *Principles*. Descartes imagines the sciences as a tree. Its roots are metaphysics, and its trunk is physics. Accordingly, metaphysics constrains physics. Physics investigates non-mental beings, in other words, bodies. Their attribute is extension, which is not distinct from space.[9] There is no space where there are no bodies. Causal efficaciousness is limited to impact: if a body moves another body, they need to touch each other. Hence Descartes goes on to formulate his impact rules (Pr II 43), in which the quantity of motion, the product of the size and the speed of the bodies, is conserved in the collision. Relatedly, he sets forth vortex-based astrophysics, which discards vacuums and explains celestial motions by positing large cycling bands which hold planets and comets in their orbits.

In Chapter 3, I explained Newton's rejection of vortex-cosmology, which is largely based on his experimentalism in the General Scholium. Here I want to focus on his earlier unpublished manuscript "De Gravitatione." Newton's disagreement with Descartes is evident right from the beginning of the manuscript. The first two definitions read: "Place is a part of space which something fills completely," and "body is that which fills place." With these two definitions, Newton advances a definition of motion as "change of place," in which rest is defined as "remaining in the same place" (Newton 2004: 13). His claims that place is a part of space and that motion is change of place assume that bodies are and move *in* space. Descartes could not accept that bodies are in space, because they *are* space. Motion cannot simply be displacement. Placement requires body-independent space. Newtonian space is something potentially absolutely empty, in which bodies are contained.[10]

Descartes goes on to distinguish between "vulgar" and "proper" concepts of motion. According to the "vulgar" conception of motion, it is "*the action by which a body passes from one place to another*" (Pr II 24). This notion is problematic as it makes a reference to displacement, and because it invokes action. The principle of inertia, as expressed in Descartes' first two laws, is in tension with an impetus-like "action" that is required to keep bodies moving. He jettisons "place" and "action" in his definition of proper motion (Janiak 2012: 405), and instead offers the following formulation:

> If, however, we consider what should be understood by movement, according to the truth of the matter rather than in accordance with common usage (in order to attribute a determinate nature to it): we can say that it is *the transference of one part of matter or of one body, from the vicinity of those bodies immediately contiguous to it and considered as at rest, into the vicinity off [some] others.*
>
> Pr II 25

Because motion is not displacement with regard to space itself, it is a thoroughly relational concept. As "transference is reciprocal," "we cannot conceive of the body *AB* being transported from the vicinity of the body *CD* without also understanding that the body *CD* is transported from the vicinity of the body" *AB* (Pr II 29). Newton's natural philosophy encompasses absolute velocity in which rest equals zero velocity, inertial motion constant velocity, and acceleration changing velocity. Bodies move with respect to the absolute space. For his part, Descartes' natural philosophy incorporates only relative motion. This implies that there is a speed difference of, say 5 mph between bodies *A* and *B*, not that *A* moves 5 mph and B stays at rest with zero velocity in comparison with a body-independent space (Slowik 2017: Section 3).

For Newton, Descartes' definitions are problematic for several reasons. Descartes thinks, or so Newton interprets, that planets are at rest as they are enclosed in vortexes. This is at odds with his note that planets tend to recede from the sun. If they travel curvilinear paths according to the heliocentric model, how could the earth, for example, truly not move (Newton 2004: 15; Janiak 2013: 405–6)? Newton (2004: 16) also points out an inconsistency in Descartes' distinction between "vulgar" and "proper" motions:

> so that the contradiction may be evident, imagine that someone sees the matter of the vortex to be at rest, and that the earth, philosophically speaking, is at rest at the same time; imagine also that at the same time someone else sees that the same matter of the vortex is moving in a circle, and that the earth, philosophically speaking, is not at rest.

As Newton explains, Descartes' definition stipulates that each body should have its own proper motion that corresponds to the nature of things. But the thoroughly relational argument Descartes makes in the *Principles* does not establish this objective: "From both of these consequences it appears further that no one motion can be said to be true, absolute and proper in preference to others, but that all—whether with respect to contiguous bodies or remote ones—are equally philosophical; and nothing more absurd than that can be imagined" (Newton 2004: 17).

Newton points out yet another inconsistency in Descartes' system (ibid.: 18). Imagine that the earth is bound by its surrounding vortex and the planets are in circular motion around the Sun, as posited by heliocentrism. Newton submits that God could stop the motion of the vortex holding the earth, which would make the earth move from rest, as the vortex itself is at rest. Newton thinks that in this case "Descartes would say that the earth is moving in a philosophical sense—on account of its translation from the vicinity of the contiguous fluid—whereas before he said it was at rest, in the same philosophical sense" (ibid.).

In this case, the motion of the earth would be initiated without an impressed force. This reveals a difficulty in Descartes' system. Descartes insists that an object should remain at rest unless there is an external cause that initiates its motion (Pr II 37). His definition of true motion does not assimilate any body's true motion with some external cause that acts on that body. From the viewpoint of Descartes' first two laws of motion, his definition of true motion is not dynamically tractable (Janiak 2013: 406). The Cartesian kinematical account, which focuses mainly on the contact of the contiguous parts of bodies, does not suffice to distinguish between relative motion and absolute motion. Descartes' first two laws become Newton's first law, but a proper causal account of motion needs to add that acceleration produced by forces is absolute motion. And hence there is a need to posit absolute space and time for an adequate analysis of different types of motion.

Next, I will proceed to Hume's philosophy of space and time. My argument propounds that Hume is clearly more Cartesian than Newtonian: he assimilates space to extension, hence casting serious doubts on the existence of a vacuum. He also focuses on perceiving change as the source of our idea of time, thus debunking absolute and universal time.

Hume on Space as Extension

In the Scholium beginning his treatment of space and time in the *Principia*, Newton argued that we should abstract from our senses and consider the true nature of space and time. In the context of formulating his philosophy of space and time, Hume (T 1.2.5 fn 12; SBN 638–9) argues for the exact opposite: "As long as we confine our speculations to *the appearances* of objects to our senses, without entering into disquisitions concerning their real nature and operations, we are safe from all difficulties, and can never be embarrass'd by any question."

Hume presents his system of space and time in the second part of the first Book of the *Treatise*. He begins his analysis by examining the "doctrine of infinite divisibility" (T 1.2.1.1; SBN 26). Hume notes that the capacity of the mind is limited. We cannot conceive an infinite amount of ideas.[11] As Hume subscribes to the early modern theory of ideas, whose objective is to reach clear and distinct ideas, such inconceivability poses a problem. When we try to think of infinities, we should be trying to think something that has an infinite number of parts. This endeavor has no bounds: it does not end at the largest number, or the smallest part. "It requires," Hume notes (T 1.2.1.2, SBN 26–7), "scarce any induction to conclude from hence, that the *idea*, which we form of any finite quality, is not infinitely divisible, but that by proper distinctions and separations we may run up this idea to inferior ones, which will be perfectly simple and indivisible."

In accordance with the copy principle, the source of any simple idea is a simple sensory impression. Hume uses the notion of a minimally sensible item. This can be, for example, a tactile colored point on a white board. Say I draw this dot on the board, walk to the far end of the classroom, and start approaching the board. The minimally sensible item is the discrete perception I have of the dot when it enters my visual field at a proper distance to it. Different people, like students in the class, have different minimally sensible perceptions, depending on each one's sensory system. Different individuals begin to perceive a minimal sensory item from different distances to the board. Also, the dot may become visible at farther distances with the aid of optical instruments, like eyeglasses. In any case, the minimally sensible points are discrete: we cannot conceive how to further divide them into parts, and there is no way to diminish them without destroying them. Taking a step back from the location in which an observer perceives the minimally sensible item will annihilate it (T 1.2.1.3, SBN 27).[12]

Although adequate, representing ideas have a threshold—to wit, the minimally sensory objects—we cannot conceive infinitesimally small or infinitely large objects. Hence no space can be divided to infinity, and no space can extend to infinity in any of its three dimensions. The latter critical point is the first clue of Hume's rejection of Newtonian infinitely large space. Hume's views are also entirely non-religious in this matter. This is different from Newton's version of the omnipresence theology, according to which God is an infinite, pure, non-corporeal extension as He is everywhere.[13]

Hume gives an example of how to acquire the idea of extension. I see a table in front of me. I acquire the idea of extension from impressions of colored points on the table. I do not see anything "further" than these visual points "disposed in a certain manner" (T 1.2.3.4; SBN 34). I also do not feel anything "further" than a three-dimensional object. The origin of our idea of extension is a cluster of visual and tactile impressions.[14] Hume unifies space and extension in roughly the same way as Descartes: The table is space; it is not in space.

There is an evident objection that could be levelled at the former claim. Hume is focused on the way we acquire the idea of space. The idea of space is just the idea of extension, provided by visual and tactile impressions. According to this view, Hume is not interested in the space physics investigates in any way. It is like he is not interested in the world, just in our ideas. They come from impressions, and where impressions come from, Hume does not know. There are at least two responses to such skepticism. First, the discussion on space and extension is not independent of philosophy of causation. I argued in Chapter 4 that Hume's concept of causation is deeply indebted to Cartesian natural philosophy. Broadly speaking, contiguity is essential to causation because impact is the causally efficacious factor (that we know of) in the collision of bodies. Physical objects change their change of motion through contact; there is no empty space or a vacuum between the bodies in the first place. Bodies do not change their states due to forces acting across vast distances in empty space, as Newton thinks, but due to contact of parts of extended matter. Hume's views are closer to the Cartesian plenist account of the universe than Newton's dynamic one. The concepts of mass (in the sense of quantity of matter) and force are crucial to Newton, but not to Descartes or Hume. Second, as I also argued in Chapter 4, causation and laws of nature are a discovered constant conjunction for Hume. It is a fact that, for example, under such and such circumstances, gunpowder explodes when set on fire. Hume thinks such a conjunction between the species of objects/events, powder and ignition, is discovered. The discovery concerns "real existence and matter of fact" (EHU 4.3; SBN 26). Hume thinks scientific inquiry is directed toward the empirical world; it is not just about our ideas. Our ideas mark epistemic restrictions: without impressions we do not have ideas and hence we could not think of the objects in question. But the centrality of the theory of ideas does not imply Hume's unwillingness to discuss physical space. We cannot know more than the empirical attributes of the world as revealed to us by sense experience. We do not know more about space than what our vision and touch convey.

There are, of course, important differences between Descartes and Hume. For Hume, we do not know the essence of body. Examining the nature of bodies is "beyond the reach of human understanding, and that we can never pretend to know body otherwise than by those external properties, which discover themselves to the senses" (T 1.2.5.26; SBN 64). Boehm (2013b: 215) notes that bodies appear to our senses. In this respect, I agree with her that Hume adopts "the standpoint of empirical natural philosophy whose fundamental tenet is, as Hume himself explains in his Introduction [to the *Treatise*], that through *experience* and *observation* we know body." Consequently, "we can attain knowledge of properties of bodies and the connections between bodies through experience" (ibid.: 216).

Hume's view is a form of relationism. A characteristic of relationism is the independence of matter thesis. Kervick (2016: 67) defines the thesis:

> The relationist defender of the independence of matter will typically hold that only the bodies themselves are physically real. They will argue that while material bodies can stand in various distance relations among themselves, and move with respect to one another, there need not also exist some extra, non-material spatial stuff that the bodies occupy, penetrate or modify in order for those distances to obtain or motions to occur.

The definition above is in accordance with Descartes' relational account of motion and in contradiction with Newton's dynamic one. There is no absolute benchmark (or we do not have its idea) for any type of motion. Motion is a comparative phenomenon, and it can be understood by considering a body moving with respect to another body (T 1.4.4.7; SBN 228):

> To begin with the examination of motion; 'tis evident this is a quality altogether inconceivable alone, and without a reference to some other object. The idea of motion necessarily supposes that of a body moving. Now what is our idea of the moving body, without which motion is incomprehensible? It must resolve itself into the idea of extension or of solidity; and consequently the reality of motion depends upon that of these other qualities.

Hume continues and adds that motion "to all appearance induces no real nor essential change on the body, but only varies its relation to other objects," and that it produces "only a difference in the position and situation of objects" (T 1.4.5.27; SBN 245–6, and T 1.4.5.29; SBN 246–7). This is clearly a relationist, non-absolutist position. Hume accounts for motion in terms of bodies moving relative to one another, without positing imperceptible space.

For Hume, we do not have an idea of a pure extension. This is something we should have if we were to believe in Newton's absolute immaterial space.

Following the general maxim, the copy principle, the senses of vision and touch convey the idea of space to the mind. Nothing appears extended, if it is not either visible or tangible (T 1.2.3.15, SBN 38–9). If the two sensible qualities of bodies are removed, there remains "only a certain unknown, inexplicable *something*" (EHU 12.16: SBN 155). "An extension, that is neither tangible nor visible, cannot possibly be conceived" (EHU 12.15; SBN 154–5); we cannot even think pure extension. Hume's position is that "we can never have reason to believe that any object exists, of which we cannot form an idea" (T 1.3.14.36; SBN 172). Boehm (2012: 94) notes the epistemic-normative character of this corollary: "If we cannot form an idea of x, we have no reason to believe in the existence of x. There is no justification for postulating entities of which we cannot form an idea." Absolute body-independent structure of space is such an entity. Newton's divine infinite space is an utterly unjustifiable piece of natural philosophy. For Hume, particular colored and tactile points make are idea of space (T 1.2.3.16, SBN 39):

> Now such as the parts are, such is the whole. If a point be not consider'd as colour'd or tangible, it can convey to us no idea; and consequently the idea of extension, which is compos'd of the ideas of these points, can never possibly exist. But if the idea of extension really can exist, as we are conscious it does, its parts must also exist; and in order to that, must be consider'd as colour'd or tangible. We have therefore no idea of space or extension, but when we regard it as an object either of our sight or feeling.

These points, or atoms or corpuscles, as he also calls them (T 1.2.3.15; SBN 38–9), are the parts that make the whole of our idea of macroscopic extension, that is, bodies. Here Hume contrasts his position to the view that a finite extension could be made from an infinite number of parts. He concludes that a finite extension cannot be infinitely divisible (T 1.2.2.2; SBN 29–30).

There is still the problem about how extension gets determined. How are spatial properties composed? In short, "*the idea of space or extension is nothing but the idea of visible or tangible points distributed in a certain order*" (T 1.2.5.1; SBN 53), as Hume encapsulates his position. According to Baxter's (2016: 173) allegory, space is extension as parts of space are next to each other like the pearls on a necklace. Jani Hakkarainen (2019) points out that this prompts a further question that should be clarified: What is the ontological status of the order of the visual and tactile points?

In Hakkarainen's view, Hume has a metaphysics of space. Hume is not merely interested in the epistemological problems related to perceiving, experiencing, or conceiving space. Hakkarainen argues that *"the ideas of extension are beings,"* which is supported by Hume's claim that "if the idea of extension really can exist, as we are conscious it does, its parts must also exist" (T 1.2.3.16; SBN 39).[15] The idea of extension not only represents extension, but it is itself extended (T 1.4.5.15; SBN 240). Moreover, Hume establishes the possibility that extension may be amenable to the idea of space: "'tis possible for extension really to exist conformable to it [the idea of extension]" (T 1.2.2.9; SBN 32).[16] The metaphysical approach implies that the existence of space is not only, even for Hume, a question for the science of human nature. There are pressing metaphysical issues that should be answered; an analysis of space (and time) is impossible without these answers.

Extension is a collection of unextended contiguous mathematical points, that are nevertheless discrete, colored, and tactile points. Unlike a physical point, a geometric point "has neither length, breadth nor depth" (T 1.2.4.9; SBN 42). As extension is constituted of simple extensionless mathematical points, space is discrete. These points are adjacent. Contiguity of the points rules out an empty space between them. This brings up Hume's doubt concerning a vacuum. Could there be a space without any visible or tangible points, an extension without any body?

Doubt on the Existence of a Vacuum

Hume's non-absolutism about space leads him to question the possibility of a vacuum. Descartes' plenism excludes vacuums in nature. They do not have any place in nature because bodies are ultimately extension. Bodies or small pieces of matter are contiguous and therefore touch each other; there is no gap or any kind of emptiness between them (Sorensen 2017: Section 10). Although Hume stands closer to Descartes than Newton on this matter, he might not eschew vacuism entirely. His critical statements are these: "it is impossible to conceive either a vacuum and extension without matter" (T 1.2.4.2; SBN 40–1), and "we can form no idea of a vacuum, or space, where there is nothing visible or tangible" (T 1.2.5.1; SBN 54). Kervick (2016: 63) argues that the relevant identifying characteristics of vacuums are the concepts of extension without matter and space devoid of anything visible and tangible. This leaves open some situation in which vacuums may exist, in a certain sense of the word "vacuum." In Kervick's (2016: 66) formulation, in addition to the independence of matter thesis, understanding Hume on space requires another relationist thesis: independence of distance. This "is the view that holding of distance relations between separated material bodies does not depend on whether or not something extended lies between the bodies."

It is important to clarify the notion of distance in this context. The classical Euclidean geometry assimilates distance and length. A line that connects points A and B is the shortest distance between the points. There might be curves connecting the points, in which case the distances between A and B are longer than the line in a Euclidean world. This basic intuitive picture of distance connects distance with extension via length. If we bluntly equate lines and curves to lengths, we miss the relational point: distance is a relation between two objects. Hume's mitigated approval of vacuums makes sense only if we consider distance as a two-place relation between bodies, not as a one-place property of the segment connecting them. "The distance relation," Kervick (2016: 67) claims, "can hold even if there is nothing extended whatsoever between A and B." Applying this definition of distance makes it possible to have relational vacuums. They are "'voids' of a certain kind, so long as these are understood in purely relational terms as systems of objects standing some distance from one another without anything at all lying between them, not even empty immaterial space" (ibid., see also fn. 14 in Kervick 2016: 83–4).

Again, none of this implies the possibility of pure extension. "We can form no idea of a vacuum, or space, where there is nothing visible or tangible," Hume argues (T 1.2.5.1, SBN 53–4). But he considers yet another case of the possibility of a vacuum, or its idea. A darkness that surrounds two luminous bodies, like two stars, does not count as a pure vacuum. If the stars were moved from an observer's visual field, the observer would see mere darkness. The idea of darkness is not however a positive idea, but a negation of the visibility of bodies. Accordingly, "'tis not from the mere removal of visible objects we receive the impression of extension without matter; and that the idea of utter darkness can never be the same with that of vacuum" (T 1.2.5.5, SBN 55–6).

Boehm notes that Hume is perfectly consistent in denying that the invisible and intangible distance is full of body without conceding that the darkness is an empty space or vacuum in the sense of pure extension without matter. The distance separating the visible bodies is "a property of objects that affects our senses in a particular manner and a capacity for receiving bodies," Boehm (2012: 91) writes. This is clearly consistent with Hume's general statement about our idea of space, which we acquire from "*visible or tangible points distributed in a certain order*" (T 1.2.5.1; SBN 53). We acquire the idea of space by considering the distance between perceivable bodies: "Upon opening my eyes, and turning them to the surrounding objects, I perceive many visible bodies; and upon shutting them again, and considering the distance betwixt these bodies, I acquire the idea of extension" (T 1.2.3.2; SBN 33). Distance is a two-property relation among bodies. Space in this sense is "known only by the manner, in which distant objects affect the senses" (T 1.2.5.17; SBN 59). Although Hume thinks space is perceivable extension, it is also a relational term to him. Perceivable objects are configured in a certain way; they can be distant from, or contiguous with, above and below each other (T 1.1.5.5; SBN 14).

From the Idea of Time to a Full-fledged Relationist Ontology

Hume's denial of infinite divisibility applies to time, too. Time is composed of indivisible parts that are parts of a succession. The parts are entirely distinct, so they do not co-exist (Garrett 2015b: 64). There are co-existing indivisible items, but then they are not parts of a succession.[17] The idea of time is "discover'd by some *perceivable* succession of changeable objects" (T 1.2.3.7, SBN 35). It is requisite that there appears a change in objects. This change can be experienced either by succession in objects, or change in their state of motion. Conceiving of time would not be possible without any "succession or change in any real existence" (T 1.2.4.2; SBN 40; see also Bardon 2007: 58). For instance, hearing five successive flute chords and abstracting the order of their succession generates the idea of time to the mind (a single ongoing chord would not be sufficient). Time consists of indivisible and finite moments which are parts of succession. An abstraction of the succession of these moments is the time we experience (Baxter 2007: 17, 22–3).

Behind Hume's reasoning are what Baxter calls existence and plurality assumptions. Only single things exist. Anything with parts is many things, not a single thing. This is reflected in Hume's concept of time: it is an abstraction of the succession we experience. No individual impression can cause the idea of time in the mind. A single object cannot have a duration; something counts as a duration only if it is a temporal complex. Moreover, Hume contends that time, or duration, can be abstracted from motions of objects. He does not accept that a steady, unchangeable object, if it is not a member of succession, could convey the idea of time. Rather, as the most important factor that needs to be satisfied for us to experience time in Hume's theory is change, motion of bodies provides change of place. In this sense, relative motion of bodies, along with succession of objects, such as auditory impressions, is a source for the idea of time.

Hume's conception of time indicates that he is a relationist. "Time is," as he puts it, "nothing but the manner, in which some real objects exist" (T 1.2.5.28;

SBN 64). Perception of time is relative to succession and/or motions of objects. Unchangeable objects, such as an ongoing chord or a pair of two motionless bodies, could not convey an idea of time to the mind. Perceiving time depends on an observer's relation to reference-objects; there is no absolute time independent of this relation. Thus there is no one absolute or universal time but many times depending on an observer's relations to reference-objects. "There is no observable evidence that the structure of time is uniform across space," Baxter (2015: 214) states. Hume asserts that we do not have an idea of time itself, independent of successive simple perceptions and relative motion: "time cannot make its appearance to the mind, either alone, or attended with a steady unchangeable object" (T 1.2.3.7; SBN 35).

We can form adequate ideas about particular objects, their dispositions, intervals, successions, and motions. We cannot form an adequate idea of space and time in themselves; we do not have the ideas of an empty space or changeless time: "'tis impossible to conceive either a vacuum and extension without matter, or a time, when there was no succession or change in any real existence" (T 1.2.4.2; SBN 39–40). Hume encapsulates his argument: "The ideas of space and time are therefore no separate or distinct ideas, but merely those of the manner or order, in which objects exist" (T 1.2.4.2; SBN 39–40, see also Wright 1983: 102).

It is quite clear from the outset that Hume is opposed to Newton's absolutism about time. For Newton, time itself is independent of any reference to change (Rynasiewicz 2014). We need to observe change to measure time, but our measuring of time is measuring relative time, not the (putative) time itself. Hume recognizes that this is both a view common to (Newtonian) natural philosophy as well as to ordinary conception of the world: "I know there are some who pretend, that the idea of duration is applicable in a proper sense to objects, which are perfectly unchangeable; and this I take to be the common opinion of philosophers as well as of the vulgar" (T 1.2.3.11, SBN 37).

The problem with "the common opinion" is that the idea of duration is always deduced from a succession of indivisible discrete objects, never from something unchangeable. Therefore, we cannot apply an idea of duration or time to anything which is perfectly invariant or motionless. In this sense, Newtonian absolute and universal time is inconceivable. It might be that there are objects that do not change in any way, but then they are not temporal. No unchangeable thing can ever have a duration (T 1.2.3.11, SBN 37). Hume's reasoning is in this regard reminiscent of a point that McTaggart (1908: 459) made more than 150 years later: "a universe in which nothing whatever changed (including the thoughts of the conscious beings in it) would be a timeless universe."[18] Nevertheless, this does not exclude the possibility of fictitious time. We do in fact in our ordinary lives fictitiously apply the idea of time to non-changing things. We believe in uniform time across space, although there is no sensory evidence for its steady flow.[19]

Relationism about Space and Time

Hume's philosophy of space and time is not first and foremost about physics. But it is related to the natural philosophical tradition. Hume's doubts about vacuums are consistent with a broadly mechanistic physics. Bodies are moved as a result of motion and contact. There is no purely empty space between the bodies, as motion and impulse are the causally efficacious parts of the process. Assimilating space to extension makes Hume side with Descartes' relationism rather than with Newton's absolute infinite space.[20] And although we believe, by means of fiction, that time flows equally from past to future, no matter what, there is no evidence for such putative absolute structure of time. As I will detail in the next chapter, Hume's critical reflection helped Einstein in part to discredit the Newtonian absolute concept of time.

7

Hume's Impact

The previous chapters of this book were devoted to natural philosophy, how it was conceived in the early modern era, and its relation to Hume. This epilogue chapter diverges from the major part of the book. I analyze Hume's partial impact on the subsequent history of physics and the philosophy of physics, while being highly selective. First, I detail the relation between Hume and Einstein's special theory of relativity (STR), and juxtapose Hume's philosophy with Einstein's philosophical analysis related to his theory. I propose that the philosopher and the physicist share an empiricist theory of concepts, and that such empiricism is apparent in the argument for the relativity of simultaneity, the key result of STR. I nevertheless argue that Humean radical empiricism is inconsistent with the ontology of STR, because the theory requires realism concerning physical events. I also treat Humeanism in the contemporary philosophy of physics. I survey the debate between necessitarian and regulationist positions, which are also called the NonHumean and the Humean metaphysics of laws of nature. The motivation of this section is to show that Hume's analyses on the topics of causation, laws of nature, and the ontology of forces are not only relics of the past, but are important in systematic philosophy as well. I then consider Hume's causal philosophy of physics in relation to subsequent acausal philosophies of physics. Lastly, I argue, by advancing the interpretations of Alex Rosenberg and Peter Millican, that Hume's position on the intelligibility of natural philosophy is interesting in the light of modern physics.

Special Relativity

On December 14, 1915, Albert Einstein wrote a letter to Moritz Schlick. The main purpose of this short letter was to compliment Schlick on his paper on special and general relativity, which Einstein had read the previous day. Einstein (1998: 161) pointed out that it was "among the best that has been written on relativity to date. From the philosophical perspective, nothing nearly as clear seems to have been written on the topic." He acknowledged that Schlick had correctly recognized that his special theory of relativity was influenced by Mach's and Hume's philosophies. Einstein wrote to Schlick, that it was "Hume, whose *Treatise of Human Nature* I had studied avidly and with admiration shortly before discovering the [special] theory of relativity. It is very possible that without these philosophical studies I would not have arrived at the solution" (ibid.).

Mach's influence on Einstein's science has been carefully perused (Holton 1968, 1992; Zahar 1977; Feyerabend 1980, 1984; Hentschel 1985; Barbour 2007; Wolters 2012). Still, Einstein insists that, more than Mach, it was Hume who enabled him to put all the decisive pieces of the puzzle together. Later, in a letter to Michele Besso in 1948, he claims: "How far (Mach's writings) influenced my own work is, to be honest, not clear to me. In so far as I can be aware, the immediate influence of D. Hume on me was great. I read him with Konrad Habicht and Solovine in Bern" (Speziali 1972: 153).

Einstein read the German translation of the *Treatise* in the early 1900s, and discussed it in the "Olympia Academy" reading group that he attended in Bern.[1] Yet his acknowledgement does not specify what it was in Hume's philosophy that he found beneficial to the formulation of STR.

Before delving into the history and philosophy of STR, there is an issue that should be tackled from the outset. The original publication of the theory, "On the Electrodynamics of Moving Bodies," was largely influenced by the nineteenth-century electrodynamic physics. As Einstein asserted retrospectively in his "What Is The Theory of Relativity" (1981c: 225), "the special theory of relativity [...] was simply a systematic development of the electrodynamics of Maxwell and Lorentz." This is consistent with the original publication. In the first paragraph of the abstract of the "Electrodynamics of the Moving Bodies," Einstein presents his famous magnet and conductor thought argument concerning Faraday's law of electromagnetic induction. This intricate argument and its important consequences for the formulation of STR have been carefully reviewed elsewhere (e.g., Earman 1989: 54; Cushing 1998: 229–30: Visser 2011: 11–15; and Norton 2010: 362–5, 2014: 83–5). To put it succinctly, the crux of the magnet and conductor argument is to argue for the principle of relativity,[2] and to argue against the ether hypothesis, or "the idea of absolute rest," as Einstein (1923: 37) says. The implication of this argument is that electric fields, as well as the dimensions of space and time, are not absolute quantities (see Norton 2014: 85). By this theoretical argument, Einstein showed that

Maxwellian-Hertzian electrodynamics, if it presupposes the ether, is inconsistent. The ether hypothesis needed to be set aside. The principle of relativity should be extended to electrodynamics. The quantities of electric field, and those of space and time, should be treated as subject to the Lorentz transformation equations.

At first sight, the original publication might indicate that STR is all about physics, math, and technology. After all, the nineteenth century saw the industrial revolution. With increasing traffic, clock synchronization became very important.[3] This is relevant for the conventionality of simultaneity, and the operationalist definition of time, which is included in the original publication of STR. The central thought experiment relates to Faraday's law, which was arguably the most important scientific principle of an industrial society: this principle is used to generate electricity. Faraday's law is encoded in Maxwell's equations, which seem to be as far as is possible from philosophical ambitions in addressing foundational epistemic, semantic, and ontic questions. The case is akin to Newton's second law and its relation to philosophy, as explicated in Chapter 1 in the context of defining natural philosophy.

My point here is not to refute the obviously true claim that STR is about physics, math, and technology. Rather, I argue that we may take a valid philosophical perspective concerning it. Although the original publication of STR ensued from a critical reflection of the nineteenth-century electromagnetic physics, the processes that led to its formulation have important philosophical roots too. This can be explained by Thomas Kuhn's (1996: 88) theory of the development of a science: scientists who introduce a new paradigm into a science work extensively in the old pre-paradigmatic science themselves. Introduction of this new paradigm, as Kuhn puts it, is "both preceded and accompanied by fundamental philosophical analyses."[4] This is no doubt the case with Einstein. He worked himself in the old pre-paradigm ether-modeled mechanics (Einstein did believe in the ether for quite a long period of time[5]), which saw electric fields and space and time as absolutes. The introduction of STR was preceded by extensive reading and discussion of philosophy, not only that of Hume's, but, among others, the philosophies of Mach, Mill, Poincaré, Spinoza, Kant, and Avenarius (see Stachel 2002: 125 and Howard 2005).

Einstein did use philosophical analysis to suit his physical ends. But just to say that he did use philosophical analysis is quite vague. What kind of philosophy? To answer this question, I quote the first chapter of his philosophically oriented text *Physics and Reality* (1981g: 283) at length. In it, Einstein clarifies his position about the need of a certain kind of philosophy to assist physics:

It has often been said, and certainly not without justification, that the man of science is a poor philosopher. Why, then, should it not be the right thing for the physicist to let the philosopher do the philosophizing? Such might indeed be the right thing at a time when the physicist believes he has at his disposal a rigid system of fundamental concepts and fundamental laws which are so well established that waves of doubt cannot reach them; but, it cannot be right at a time when the very foundations of physics itself have become problematic as they are now. At a time like the present, when experience forces us to seek a newer and more solid foundation, the physicist cannot simply surrender to the philosopher the critical contemplation of the theoretical foundations; for, he himself knows best, and feels more surely where the shoe pinches. In looking for a new foundation, he must try to make clear in his own mind just how far the concepts which he uses are justified, and are necessities.

In the quote above, Einstein suggests that "when looking for a new foundation," the physicist needs to make clear "just how far the concepts which he uses are justified, and are necessities." This indicates that he is essentially interested in epistemological problems. He did value epistemological analysis in his physical research when he was forming a novel physical theory, such as STR (see also Einstein 1949b: 683–4). It should be noted that he did not regard himself as a "systematic epistemologist," but rather an "unscrupulous opportunist." Still, his "occasional utterances of epistemological content" can be interpreted as having systematicity, as he himself also allows this in his summary to the volume of *Library of Living Philosophers* (ibid.: 683). Interestingly, Hume's *Treatise*, a work that Einstein reported to have studied with enthusiasm, begins with asking essentially the same kind of questions that Einstein dealt with in his more philosophically oriented texts: what are the origin, meaning, and justification of our ideas?[6]

As John D. Norton (2010) has argued, what Einstein got from Hume (and Mach) was an empiricist theory of concepts. Norton's main thesis is that the young Einstein was most influenced by the way Hume saw concepts to be grounded in sensory impressions. The early Einstein implemented empiricism in his argument for the relativity of simultaneity. If the concept of simultaneity is grounded in sensible impressions, such as in visual sensations of immediate light flashes in two mirrors, it follows (given the invariant one-way speed of light) that there is no absolute simultaneity. Different inertial reference frames can observe the timely order of two spatially distant events, the two light flashes, in different order. The revision of the concept of simultaneity defied the absolute Newtonian and the *a priori* Kantian character of time. As Einstein recognized later in his

1949 autobiographical writing: "The type of critical reasoning required for the discovery of this central point [the denial of absolute time, or simultaneity] was decisively furthered, in my case, especially by the reading of David Hume's and Ernst Mach's philosophical writings" (Einstein 1949a: 53). In Norton's interpretation, this "critical reasoning" eventually helped Einstein to reconcile the two postulates of STR, namely the invariance of laws of nature and the constancy of speed of light in a vacuum.

In the next section, I wish to take a closer look at the implementation of concept empiricism into the argument for the relativity of simultaneity. I will first argue that Einstein indeed assumes an empiricist theory of concepts, hence contrasting his position to Newton's absolutist and Kant's transcendental arguments. In this sense his position is Humean. But what I also wish to show is that STR requires a realist commitment to events. I will explain how this makes Einstein diverge from Hume's radical concept empiricism.

Concept Empiricism and the Relativity of Simultaneity

As noted in the previous chapter, for Newton time itself is imperceptible. Our measures of time are relative. The more accurate the clock the better it imitates absolute time, which flows equably. Temporal order is grounded in time itself; all observers agree whether two events are simultaneous or not. Another highly influential pre-relativity view on the concept of time is present in Kant's first *Critique*. Kant is against Newton's view: "time is not something that would subsist for itself" (KdRV, B49/A33). Kant argues that time is an *a priori* form of sensibility. In the second section of "Transcendental Aesthetics," he writes that:

> Time is not an empirical concept that is somehow drawn from an experience. For simultaneity and succession would not themselves come into perception if the representation of time did not ground them *a priori*. Only under its presupposition can one represent that several things exist at one and the same time (simultaneously) or in different times (successively) [...] Time is therefore given *a priori*.
>
> KdRV, A30/B46

Einstein's empiricist argument on the concept of time defies both the Newtonian and the Kantian positions. Einstein (2001: 27) asks us to imagine two observers, M and M'. M is at rest on an embankment and M' is moving in a train from left to right with constant velocity:

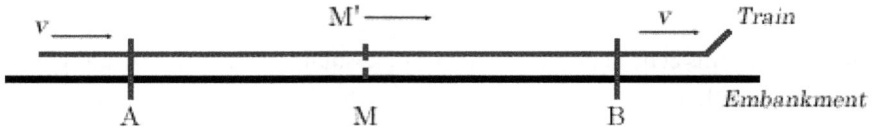

Figure 4 Relativity of simultaneity.

Both observers have two mirrors that are inclined at 90 degrees. These enable them to see the light coming from lightning bolts that strike points A and B at the embankment. In the frame of M, the strikes are simultaneous. The distances from A to M and from B to M are equal. Light travels with constant velocity, independently of direction. So when M sees the flashes simultaneously, it must be that they happen simultaneously. M' is moving toward B and away from A. Hence she sees the lightning strikes successively. In her frame the strikes are not simultaneous. For her, B happens before A, because the distances from A to M' and from B to M' are equal. It must be that for her the physical event itself, the lightning striking point B, takes place before the other event, lighting striking point A. If there were a third observer M" moving from right to left, she would conclude that in her frame the bolt hitting A happens before the one hitting B. The result:

> Events which are simultaneous with reference to the embankment are not simultaneous with respect to the train, and vice versa (relativity of simultaneity). Every reference-body (coordinate system) has its own particular time; unless we are told the reference-body to which the statement of time refers, there is no meaning in a statement of the time of an event.
>
> Einstein 2001: 28–9

Importantly, Einstein renders the concept of time empirical. Observations of temporal order are needed to justify our notions of simultaneity. "The definition of simultaneity [...] must supply us with an empirical decision as to whether or not the conception that has to be defined is fulfilled," he writes (ibid.: 25).[7]

An empiricist argument like this is in tension with both the absolutist and the transcendental arguments on the concept of time. Newton's absolute simultaneity or succession is grounded in the unobservable flow of time. It is obvious that in this regard relativity and Newton's dynamics are incompatible. STR's Lorentz transformations establish the relativity of clock time,[8] and the argument for the relativity of simultaneity is inconsistent with the notion of time itself. There is no absolute fact to the matter of whether two spacelike separated events are simultaneous. In Einstein's (1981e: 267) view, the "tremendous success" of Newton's dynamics "prevented him and the physicists of the eighteenth and nineteenth centuries from recognizing the fictitious character of the foundations of his system."

Moreover, the empiricist argument is inconsistent with the transcendental argument.[9] For Kant, time is grounded in the *a priori* forms of human intuition. In his *Meaning of Relativity*, Einstein contends that "I am convinced that the philosophers have had a harmful effect upon the progress of scientific thinking in removing certain fundamental concepts from the domain of empiricism" (Einstein 2003: 2). When referring to "the philosophers [...] removing certain fundamental concepts from the domain of empiricism," he does not identify any philosopher by name. Since in this context he targets his critique at the *a priori* status of space and time, it is reasonable to assume that he is criticizing Kant's transcendental idealism about space and time. In Einstein's view, time is an empirical concept. Simultaneity or succession is not given *a priori*.[10] To put the concept of simultaneity into a more "malleable" form, Einstein went on to "purge" the *a priori* elements from concepts (see section 3.2 of Norton 2010). In his correspondence with Max Born in 1918, Einstein writes that he wishes to "water down the '*a priori*' to 'conventional'" (Howard 1994: 52).

Einstein does not think that the source of concepts is impressions, as Hume would (Lenzen 1949: 360). Rather, their origins are in the free creations of the human mind. In this sense, Einstein can be seen to be closer to Kant than Hume. Kant's philosophy emphasizes the active and constructive aspects of perception, whereas Hume's understanding of perception is that it is passive; Hume thinks that senses are inlets through which ideas are conveyed (EHU 2.7; SBN 20, EHU 12.9; SBN 152). Nevertheless, Einstein's rhetoric in his correspondence with Born and Paul Ehrenfest indicates that he is more sympathetic to Humean empiricism than Kantian transcendentalism. He claims that "the details [of Kant's philosophy] do not fit,"[11] and that "it is not as good as his predecessor's Hume's work" (Howard 1994: 52). In 1916, he wrote to Ehrenfest that "Hume really made a powerful impact on me. Compared to him, Kant seems to me truly weak" (ibid.: 50). In Hume and Einstein, space and time are empirical ideas and concepts. They are not *a priori* preconditions for all possible experience, as Kant thinks.

Mara Beller (2000) has questioned the relevance of Hume's philosophy to Einstein's empiricism. Beller refers to Einstein's article "Remarks on Bertrand Russell's Theory of Knowledge," which she takes to provide evidence against the empiricist reading. She writes the following:

While Einstein's references to Hume's impact are often treated as a token of Einstein's empiricism, the discussion in this paper leads to a different appreciation of Hume's role in Einstein's critical thinking: "Man has an intense desire for assured knowledge. That is why Hume's clear message seemed crushing: the sensory raw material, the only source of our knowledge, through habit may lead us to belief and expectation but not to the knowledge and still less to the understanding of lawful relations" (Einstein 1981j: 32). Hume's impact on Einstein is then not necessarily an influence in the direction of empiricism, as usually assumed.

Beller 2000: fn. 5, 102–3

I am partially sympathetic to Beller's reading. Einstein distances himself from the view that "the sensory raw material" would be "the only source of our knowledge." According to his view, concepts require conventions to be applicable. This is not the case with Hume. However, the passage quoted by Beller does not prove that "Hume's impact on Einstein is then not necessarily an influence in the direction of empiricism," nor does it even prove the weaker claim that there are no similarities between Hume's and Einstein's empiricisms. In the same text, "Remarks on Bertrand Russell's Theory of Knowledge," Einstein (1981j: 33) supports an empiricist theory of concepts, as he writes that "all thought acquires material content only through its relationship with that sensory material." The quotation that Beller has selected concerns Einstein's conception of the knowledge and of the understanding of lawful relations. She is correct that Einstein did not support a Humean understanding of our belief and expectation of lawful relations as being founded on custom and habit. Yet this does not provide evidence that Einstein did not subscribe to an empiricist understanding of *concepts*.

So far, I have argued that Einstein assumed an empiricist theory of concepts, which was important for making the concept of time an empirical one. Such an argument partly privileges concept empiricism over absolutist and transcendental arguments on time's concept. But Humean radical empiricism also entails a problematic position. It is in tension with a key ontological requirement of special relativity, to wit, the reality of physical events.

The special theory of relativity assumes the following metaphysical view: There is a distinction between true events and our observations of, say, light flashes perceived by the observer. To make this assumption clear, and to show the tension it inflicts on Hume's philosophy, we may consider the frame of M in Figure 4. In the frame, M is located at the midpoint between points A and B on the embankment. There might be more experimenters in the same frame. For example, observer N is located just next to point A, and observer O just next to point B. In Einstein's example, M sees the lightning strikes simultaneously. But N sees the light coming from A first, and from B after. O sees the flashes in inverse order than N. Which one of the three observers sees the temporal order correctly? M, because she sees the strikes in the order in which they really happen. To clarify this point, it is useful to quote from a physics textbook (Knight 2008: 1153): "Simultaneity is determined by when the events actually happen, not when they are seen or observed. In general, simultaneous events are *not* seen at the same time because of the difference in light travel times from the events to an experimenter."

It should be added that "when the events actually happen" is relative to an inertial reference frame. "But once a frame is chosen," to quote from Bradley Dowden (2018), "this fixes objectively the time order of any pair of events." In a particular frame, there is a fact to the matter of whether two events happen at the same time or not.[12]

There are also cases in which all observers agree on the temporal order of events. All observers can agree on the spacetime intervals of events. Hermann Minkowski worked out the mathematics of spacetime in his 1908 article "Space and Time" in support of Einstein's STR. Here it is noteworthy to reproduce the ingress of the article in its entirety (Minkowski 1923: 75):

> The views of space and time which I wish to lay before you have sprung from the soil of experimental physics, and therein lies their strength. They are radical. Henceforth space by itself, and time by itself, are doomed to fade away into mere shadows, and only a kind of union of the two will preserve an independent reality.

Inertial observers moving with respect to each other measure temporal and spatial intervals differently—if one observer measures a shorter temporal interval and a larger spatial interval, the other observer measures a longer temporal interval and a smaller spatial interval—but the spatiotemporal interval between two events is invariant (Bardon 2013: 69). Minkowski's (1923: 84) light cones picture this idea (Figure 5).

Relativity of simultaneity applies only to spacelike separated events that fall outside the cones. If any signal traveling at the speed of light or slower than it from an earlier event reaches the later event, the order of the events is invariant. Hence an observer may have an absolute past and an absolute future.

Why is this relevant for assessing the relationship between Humean philosophy and STR? Because STR requires that events happen independently

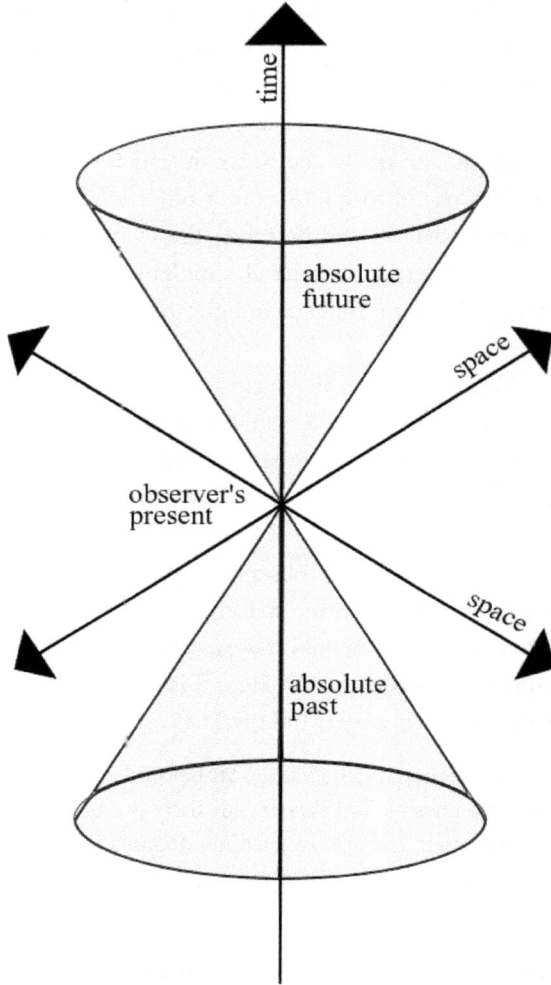

Figure 5 Absolute past and absolute future in an observer's Minkowski light cones.

and before the observer sees them. This is in tension with Hume's radical empiricism as expressed by his copy principle. The distinction between "what really happens" and "how we see things happening" fits poorly with Hume's radical empiricism. But this is exactly what Einstein's argument for the relativity of simultaneity maintains. It is requisite that, as Einstein (1936: 358) himself points out, we "differentiate between 'simultaneously seen' and 'simultaneously happening." One could say that the fundamental entity of STR is a real event; that is, some physical activity taking place at a certain spatial point at a certain time. It is not an immediate particular perception of an object.

On the other hand, there may be a way in which Hume's copy principle could be made consistent with Einstein's argument. There might be a frame in which an observer sees the lightning flashes simultaneously, but hears the strikes successively. One can infer the non-simultaneity of the lightning strikes because the strikes make successive auditory impressions. Could this enable Hume to say that physical events are different from our perceptions, and that different visual and auditory perceptions have a common natural origin? There are some explicit passages in the *Treatise* which indicate that Hume thought that our perceptions are affected by physical causes. For example, in T 2.1.1.2 (SBN 275), Hume explicitly states that perceptions in the mind "depend upon natural and physical causes." If our visual and auditory perceptions have such natural causal origin, then Hume's position would readily be reconcilable with the argument for the relativity of simultaneity.

STR is partly compatible with a common-sense realism. We can think of light waves as messengers. In principle, light carries information in the same way as other messengers, like radio waves delivering the news and a mail carrier passing on a letter. Common sense tells us that the events that are informed are independent and prior to the receivers acknowledging them. To amplify this commonsensical point, take a castaway on a deserted island. All of a sudden, the castaway notices a ship passing by the island. Fortunately, she is well equipped and has some means of contacting the ship; she has in her possession a flare, a loudspeaker, a piece of paper, and a bottle. The castaway can send the ship a message by either light, sound, or a piece of paper. Using simple common sense, we can affirm that the event (in this case, the castaway's unwanted circumstances and her need for help) that is informed to the receiver (the rescuing ship) occurred prior and independent of it being received.[13]

In many instances, STR is at odds with common sense, whereas the Newtonian picture is closer to a commonsensical understanding. Time dilation and length contraction are, from the viewpoint of everyday experience and reasoning, truly astonishing. But not everything in STR is mindboggling, and common-sense realism should not be abandoned without forcefully persuasive reasons. Of course, common-sense realism should not be just dogmatically accepted, either. One needs to address the skeptical challenge. In Keith Lehrer's (1978: 358-9) formulation, "before skepticism can be rejected as unjustified," common-sense realism still needs to satisfy the following conditions:

> [S]ome argument must be given to show that the infamous hypotheses employed by the skeptics are incorrect and the beliefs of common sense have

the truth on their side. If this is not done, then the beliefs of common sense are not completely justified because conflicting skeptical hypotheses have not been shown to be unjustified. From this premise it follows in a single step that we do not know these beliefs to be true because they are not completely justified.

To follow Noah Lemos' (2010: 59) analysis of the above-mentioned requirements, the common-sense realist needs to establish a positive argument in favor of realism and show that any skeptical argument is in contradiction with the positive argument. These requirements can be satisfied among experimenters in the same inertial frame, as there is a univocal truth to the matter in which temporal order events happen. It must be that the events take place independent and prior to their observations, because they do not necessarily observe the temporal order in the order in which the events take place. So there is a positive argument, true time ordering of events, which excludes all other logically possible time orderings.

The aspects of STR which are not in accordance with common sense are well established, so they should be accepted. But it is in accordance with STR that perception-independent events exist in their own right and flood our perceptions with light-mediating information. This is, in essence, common-sense realism. It seems Hume argues against even a very commonsensical form of realism.[14] A strikingly skeptical argument appears in T 1.3.5.2 (SBN 84):

> As to those impressions, which arise from the senses, their ultimate cause is, in my opinion, perfectly inexplicable by human reason, and 'twill always be impossible to decide with certainty, whether they arise immediately from the object, or are produc'd by the creative power of the mind, or are deriv'd from the author of our being. Nor is such a question any way material to our present purpose. We may draw inferences from the coherence of our perceptions, whether they be true or false; whether they represent nature justly, or be mere illusions of the senses.

Here Hume says that our perceptions may be caused by mind-independent nature, by our minds themselves, or by God. This is a thoroughly skeptical remark: there is no intelligible way to differentiate between the three (or perhaps even more) options. Hume is no Lockean representational realist.

There is an additional reason to think that Hume's radical empiricism is not fully in accordance with the argument for the relativity of simultaneity. An observer could see the flashes of light simultaneously, but hear the rumbles of thunder successively (see Slavov 2016b: 36–7). Hume could not infer that this observer makes the right conclusion because different senses are distinct. The copy principle does not countenance the common natural origin of the five different sensory impressions.

Although Hume's philosophy is of central importance for understanding the philosophical background assumptions of the concepts of space and time in STR, Einstein's argument for the relativity of simultaneity includes assumptions which are incompatible with the old Humean radical empiricist interpretation of the copy principle. To emphasize the point, agreeing with STR requires a realist ontology of physical events. It is unclear whether the radically empiricist and skeptical aspects of Hume's philosophy could be made consistent with such realism.

Humeanism and NonHumeanism about Laws of Nature

Hume's analysis of the laws of nature has been influential in subsequent accounts of the metaphysics of laws. In the contemporary philosophy of physics, it is commonplace to introduce two rival positions on the laws of nature: the Humean and the NonHumean positions. The former position maintains that laws are records of universal generalizations which do not instantiate necessity. The latter states that laws govern and necessitate the behavior of objects.[15] The opposing views are not the only alternatives. Tuomas Tahko (2015: 513) defends a middle ground between the two conceptions. According to his hybrid view, "some laws are contingent and some laws are necessary."

The NonHumeans[16] (for example, Dretske 1977; Tooley 1977; Armstrong 1983; Carroll 1990, 1994, 2012) argue that laws govern the behavior of objects and events. Laws "say" what must happen, and what cannot happen. Carroll (2012) begins his analysis by recognizing that laws are generalizations. Still, not all generalizations are laws. In his view, laws of nature can be distinguished from non-laws in that laws instantiate some sort of necessity. Laws of nature are restricted by physical, not logical, necessity, whereas non-laws are not. To illustrate his point, Carroll (2012) refers to Bas C. van Fraassen's example of the diameter of gold and uranium spheres in his work *Laws and Symmetry* (1989: 27):

Consider the unrestricted generalization that all gold spheres are less than one mile in diameter. There are no gold spheres that size [bigger than a mile in diameter] and in all likelihood there never will be, but this is still not a law. [. . .]

The perplexing nature of the puzzle is clearly revealed when the gold-sphere generalization is paired with a remarkably similar generalization about uranium spheres:

> All gold spheres are less than a mile in diameter.
> All uranium spheres are less than a mile in diameter.

Though the former is not a law, the latter arguably is. The latter is not nearly so accidental as the first, since uranium's critical mass is such as to guarantee that such a large sphere will never exist.

In Carroll's (1990: 185) formulation, laws of nature have a specific "modal character, a modal character not shared by accidentally true generalizations." This modal character is physical necessity, which excludes physical impossibilities that a given law restricts. It is physically impossible that there would be uranium spheres larger than one mile in diameter, as uranium's critical mass prevents this. This is necessitated by the laws of nuclear physics.

Carroll's approach can be generalized to cover many of the laws of nature, notably those of classical physics. For instance, it is physically necessary that if a net force $\vec{F}$ is exerted on an object, the object will accelerate in the direction of $\vec{F}$. It would be physically impossible for this object to remain in its initial state of motion or to move to some other direction than $\vec{F}$. The same holds for the laws of electricity: if a proton comes to an electric field $\vec{E}$, it would be physically impossible for it not to accelerate into the direction of $\vec{E}$.

At first sight, it might seem that probabilistic quantum mechanics is non-necessitarian.[17] The nonHumean does not have to agree. In the words of Stephen Hawking and Leonard Mlodinow (2010: 72):

> Quantum physics might seem to undermine the idea that nature is governed by laws, but that is not the case. Instead it leads us to accept a new form of determinism: Given the state of a system at some time, the laws of nature *determine* the probabilities of various futures and pasts rather than determining the future and past with certainty.

For example, when I roll a die, I cannot beforehand determine whether the outcome is 1, 2, 3, 4, 5, or 6. There is no law that determines the specific outcome. The prediction is based on a probability. According to the governing conception of laws, however, the laws of probability determine the probability for each side of the die facing up to be $\frac{1}{6}$. Although the specific outcome is not necessitated by any law, the law still determines the probability of the outcome.

The Humeans (for example, Lewis 1973; Ramsey 1978; Swartz 1995; Beebee 2000) disagree with the idea that there are laws that necessitate the behavior of objects. Laws are nothing over and above the regular patterns of the universe. There is nothing extrinsic to objects and events that could be called a law. Necessitarian views lead to a dubious metaphysics. If laws are external to physical processes, and if laws govern their behavior, then there should be a non-physical level of reality more fundamental than the physical.[18] To come back to the example of the die, it is as if external to it (a physical object, a piece of plastic), there is something that makes it move the way it does. Such a view seems metaphysically suspect: there should be something non-physical, the law, which determines the physical, the object, to behave the way it does.

Humeans may distinguish laws and accidentally true generalizations. For example, it is a true generalization that all the beer pint glasses in my kitchen cupboard are currently empty. This is clearly not a law of nature. The Humean (or Ramsey-Lewis) account explains this by pointing out that a "No beer pints in kitchen cupboard" generalization does not play a role in any axiomatic system (Beebee 2000: 575–6). It is not an axiom from which one could deduce other physical propositions with any predictive power. Although the Ramsey-Lewis account argues for counterfactual dependencies and even physical necessity (ibid.), and hence deviates from Hume's views about laws, it is Humean in the sense that it offers a descriptive notion of a law of nature. Beebee (ibid.: 578) explains the difference by emphasizing the separate takes Humeans and NonHumeans have on determinism:

> One way of bringing this out is to consider the thesis of determinism. We can characterize determinism in the following rough and ready way: the state of the universe at any given time together with the laws of nature *determines* what the state of the universe will be at any future time. But what does "determines" mean here? For the Humean, the laws and current facts determine the future facts in a purely logical way: you can *deduce* future facts from current facts plus the laws. And this is just because laws *are*, in part, facts about the future. So for the Humean, the notion of determination is, as it were, a metaphysically thin one. This contrasts sharply, I think, with the notion of determination which the anti-Humean has. For the anti-Humean, the notion of determination is a metaphysically meaty one. It isn't just that the laws plus current facts *entail* future facts; rather the laws 'make' the future facts be the way they will be: the laws are the ontological *ground* of the future facts.

To apply the notion of Humean determinism to causal power, we can say that there is no causal power (or we do not have its idea) which determines the future course of the world. By knowing the relevant initial conditions and lawful regularities, we may predict the future outcome of events. But the laws are not grounded on some deeper level causal power that brings about their effects. Instead, laws are constant conjunctions for which we know of no exceptions. Hume himself puts the point as follows (EHU 6.4; SBN 57–9):

> There are some causes, which are entirely uniform and constant in producing a particular effect; and no instance has ever yet been found of any failure or irregularity in their operation. Fire has always burned, and water suffocated every human creature: The production of motion by impulse and gravity is an universal law, which has hitherto admitted of no exception.

"Humeans," Beebee (ibid : 580) concludes, "do not require laws to 'do' anything: like accidentally true generalizations, laws are at bottom merely true descriptions of what goes on. Thus for the Humean there is no need for any ontological distinction between laws and accidents."

Causation and Laws

In Chapter 4, I argued that Hume's concept of causation is in tension with Newtonian dynamics. Hume thinks that 1) causes and effects are contiguous to each other, 2) causes temporally precede effects, and 3) causes and effects are distinctly separable (T 1.3.15.3–4; SBN 173, EHU 4.11; SBN 30). The problem with the three preconditions is that the laws of dynamics and the law of conservation of momentum are not consistent with such preconditions. As the force of gravity is a long-range force, the first criterion is not satisfied.[19] Two objects exert a mutual force on each other, although there is nothing in between them. Second, the exercise of force is simultaneous with acceleration. There is no temporal succession in dynamical interactions, as the second precondition demands. Third, forces are not individual things or properties of bodies. Rather they "emerge" (it is not clear what would be the appropriate verb to use, since the application of active words may be misleading here) from interactions of at least two bodies. For example, when I push the table in front of me with my hand, the table "pushes" (here the application of such verbs might lead to anthropomorphizing nature) my hand back with equal and opposite force. Is it possible to distinguish between cause and effect in this scenario? Is my pushing the cause, and the opposite force of the table on my hand the effect, or vice versa?[20]

The same problem can be observed when applying the notion of causation to the law of conservation of momentum, which Hume takes to be a discovered causal law (EHU 4.13; SBN 31). The law states that in an isolated system, the total linear momentum, the quantity of motion, of the system is a conserved quantity. When a cue ball is shot and hits the object ball in a game of pool, the total momentum is preserved in the impact. Can we differentiate cause and effect in this scenario? Hume seems to think that the quantity of motion of the cue ball is the cause, and the change of motion of the object ball is the effect. However, the resultant motion of the object ball does not suffice to show that it is the effect in the scenario; both of these objects may change their state of motion. It is perfectly uncontroversial to say that the quantity of motion is conserved in the impact of the balls, since this is what the proposition concerning the law actually states. But it is highly controversial to say that one of the objects is the cause for the other object's motion, as Hume assumes. The law does not say that momentum is "transferred" from one body to another; the law says that momentum is conserved in the system of the bodies.

If the causal interpretation of laws is nevertheless considered to be a desirable goal, it might be that the projectivist interpretation (or Coventry's quasi-realism) could surmount this conflict. According to Beebee's (2007: 225) projectivist interpretation, "Hume holds that our causal thought and talk is an expression of our habits of inference. On observing *a*, we infer that *b* will follow, and we 'project' that inference onto the world." To follow this line of interpretation, causation is not something discovered in nature. When we say that *a* causes *b*— for example, that the impact of the cue ball (*a*) causes the object ball to change its state of motion (*b*)[21]—we are merely projecting our habits of inference onto natural phenomena. Projectivism, or quasi-realism about causation, stresses that we should focus on the aspects of the human mind, or human nature, which make us think and say that objects or events are causally related. Although this reading seems plausible for reconciling Hume's causal philosophy with causal interpretation of laws, there are still some caveats that would have to be taken into account. Especially in the first *Enquiry*, Hume seems to think that causal relations are discovered in nature by experience; they are not mere projections.

The preceding reasoning indicates that Hume's view of causation is inconsistent with classical Newtonian physics (although it is more hospitable to Cartesian physics). This is not to say that there could not be a causal interpretation of it. Newton himself argued for a counterfactual condition of causation. This is apparent in his formulation and the first explaining sentence of the first law of motion in the Axioms, or Laws of Motion in the *Principia*:

Every body perseveres in its state of being at rest or of moving uniformly straight forward except insofar as it is compelled to change its state by forces impressed. Projectiles persevere in their motions, except insofar as they are retarded by the resistance of the air and are impelled downward by the force of gravity.

The counterfactual theory of causation, which Hume briefly mentions,[22] does not include a reference to contiguity or temporal priority. It is therefore consistent with Newton's physics. Another compatible option is the interventionist condition for causation.[23] The following scenario in Newtonian physics may be explained with a manipulability theory of causation. Consider the ideal-gas law, $pV = nRT$. This law states the interrelationship between four state variables, pressure, p, volume, V, the moles of the gas, n, and the temperature of the gas, T (R is the universal gas constant). Take an isolated system, a sealed tank of gas. Then work is done on the system: one side of the tank is pressed and the volume decreases. Consequently, the pressure and temperature inside the tank will rise. If we were to just examine the state variables of the law, we could not identify a cause and an effect. But if the system is manipulated, a change occurs within the system. Therefore, intervention on the system denotes the cause, and this intervention is what brings about the effect, the increasing temperature and pressure.

It should be remarked, however, that the causal philosophy of physics has been challenged by philosophers like Berkeley (1992), Comte (2012), Mach (1919), Russell (1953), Waismann (2011), Field (2003), Norton (2003), and Ladyman and Ross (2007).[24] It might be that Hume, and many other early moderns such as Descartes and Newton, are wrong in assuming that laws of nature are causal. Maybe laws should not be understood in terms of causes and effects at all? Perhaps it is wrong to assume that there are causes to be discovered in the first place?

Comte argued that causation belongs to the metaphysical age. Saying that "force causes acceleration" is analogous to people in the theological age saying that the "wind blows" or the "thunder strikes." Causation is an anthropomorphic projection of agency onto the world, which should be eliminated in the final positive stage. Mach and Russell argued that Newtonian physics can be written with differential equations, and hence the term "force" may be eliminated from physics. In his classical article "On the Notion of Cause," Russell (1953: 395) claims, by using Newton's law of universal gravitation as an example, that "in the motions of mutually gravitating bodies, there is nothing that can be called a cause, and nothing that can be called an effect; there is merely a formula." Rather than interpreting the law in causal terms, Russell proposes that a physical system can be expressed in terms of differential equations, which render the configuration of particles theoretically calculable. "That is to say," he continues, "the configuration at any instant is a function of that instant and the configurations at two given instants. This statement holds throughout physics, and not only in the special case of gravitation. But there is nothing that could be properly called 'cause' and nothing that could be properly called 'effect' in such a system."

In contemporary philosophy of physics and naturalistic metaphysics, Norton (2003) and Ladyman and Ross (2007) make roughly the same point as Russell: causation is a folk concept, not something that is discovered in fundamental physics (however, the concept is used in many other disciplines).

Physics and Intelligibility

As I argued in Chapters 2 and 3, Hume denies that there is reason or intelligibility in causal relations. Predication of qualities is not founded on reason. Instead, it is founded on causation, which is founded on experience. This is important in the context of the contemporaneous result of physics, namely the law of universal gravitation. There is no reason why gravity pertains to bodies with mass as Leibniz requires. The way gravity causes changes of motion is unintelligible. Thus Millican (2007: xxix) puts it: Hume undermines "even the *ideal* of causal intelligibility." Rosenberg (1993: 73) agrees: "[T]he whole notion that causation rests on or reflects the intelligibility or rationality of sequences among events is a mistake. Accordingly, for Hume, the aim of science cannot be to reveal the intelligible character of the universe, but simply to catalogue the regularities that causal sequences reflect." An utterly strange effect might follow a common cause—"The falling of a pebble may, for all we know, extinguish the sun; or the wish of a man control the planets in their orbits" (EHU 12.29; SBN 164)—because causation is a regular, not an intelligible, relation.

I think Hume does not entirely eschew intelligibility because his concept of causation is in many ways embedded in the mechanistic, Cartesian and Leibnizian traditions of natural philosophy.[25] Hume thinks that causation does not make sense without temporal succession.[26] I wish to omit this point for the purpose of this section, because here I am not dealing with causation at all. I will focus on Hume's rejection of intelligibility, something which can be taken seriously considering the results of modern physics. This point is mentioned by Millican (2007: ix–x): "This outlook, revolutionary in its time, was to be powerfully vindicated during the twentieth century as the successes of relativity theory and quantum mechanics forced scientists—often very reluctantly—to accept that intuitive 'unintelligibility' to human reason is no impediment to empirical truth."

Although Millican very clearly refers to unintelligibility of modern physics, he does not elaborate on the point. To my knowledge, this point has not been taken up by anyone else, either. I shall remedy this omission by considering the unintelligibility of time dilation in special relativity and the double-slit experiment in quantum mechanics. My objective is not to defend the Humean view. Rather, I show how Hume's view, or the Humean view, is truly fascinating in assessing the philosophical ramifications of some central physical results.

Humean Outlook on Time Dilation and the Double-slit Experiment

Some things are clearly obvious. As if the light of reason shows that this is the way things are, and, more importantly, this is the way things cannot be. It seems intuitively clear and reasonable that a parent cannot be younger than her child, and that particles cannot be waves. But consider the examples below.

Matt, aged 30, is the father of Pat, aged 2. Matt goes on a trip in a spaceship that travels at 0.99c. Pat stays on earth. In Matt's timeframe, his trip takes 10 years. When Matt returns to earth, he is 40. With a Lorentz time dilation equation, $t = \mu t'$, in which t is the proper time, μ is constant $\sqrt{1 - \dfrac{v^2}{c^2}}$, and t' is the dilated time, we can calculate Pat's age when Matt gets back:

$$t_{Pat} = \frac{t_{Matt}}{\mu} = \frac{10}{\sqrt{1 - 0.99^2}} \approx 70 \; years.$$

A son can be roughly 30 years older than his father.[27]

Then the quantum example. In an experimental setup, it is possible to shoot individual particles, like electrons, toward two slits, 1 and 2, separated by a very small distance. Behind the slits there is a screen to which the electrons collide and leave a mark. Classical, commonsensical understanding tells us that the electrons go through either slit 1 or 2, and form particle-like marks on the screen. The marks should appear on the screen behind slit 1 or 2: behind 1 if the particle goes through slit 1, and behind 2 if it goes through 2. When enough particles are shot, however, the pattern looks like this (Figure 6).

The image on the detector screen indicates a wave-like motion. The inference pattern with its varying intensities is consistent with the predictions of the wave-

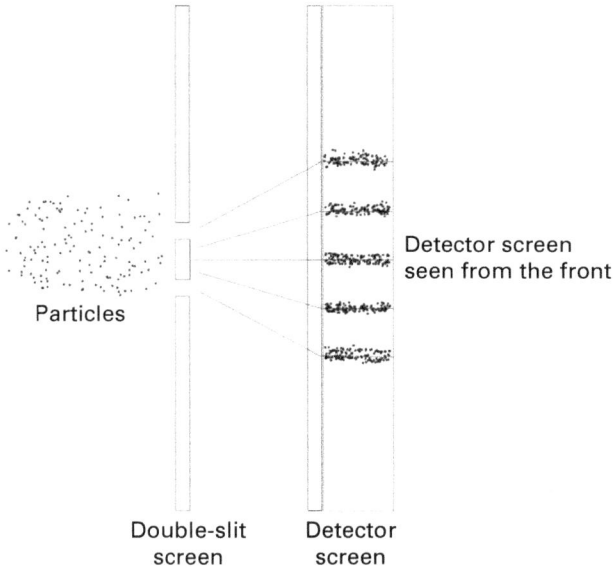

Figure 6 Structured interference pattern on the detector screen.

mechanics. "Electrons arriving one by one is particlelike behaviour; the resulting collective interference pattern is wavelike behaviour," comments John Polkinghorne (2002: 23). Such particle–wave duality is deeply counterintuitive. If something is a particle, it is distinctly located in space; if something is a wave, it is spread out across space. In the words of Marianne Freiberger (2012): "It's a very weird result indeed but one that has been replicated many times—we simply have to accept that this is the way the world works."

We can see how Hume's denial of intelligibility in tandem with his support for experimentalism is relevant considering current physical knowledge. To deny that a child is always younger than his parent, or to deny that particles and waves are mutually exclusive, is to accept something astonishingly, completely bizarre. As Rosenberg (1993) and Millican (2007) have called it in their Hume scholarships, there is something unintelligible about the findings of science. Unintelligibility does not entail contradiction; there is nothing contradictory in the examples I have given. But what matters is that there is an immense amount of experimental evidence for such unintelligible views. We get the certainty of the results by repeating them countless times. For Hume, we have an inductive proof for such theories, because there are no exceptions to them (EHU 10.4; SBN 110–1; see also fn. 10; SBN 56, and 10.12; SBN 114–15): "A wise man, therefore, proportions his belief to the evidence. In such conclusions as are founded on an infallible experience, he expects the event with the last degree of assurance, and regards his past experience as a full *proof* of the future existence of that event." The outcomes of these experiments are essential, even though they are completely weird. Nature may be in many ways utterly strange. What matters for our knowledge of nature is repetition and reproducibility, not putative reason or intelligibility.

Notes

Preface

1   According to a Philpapers.org survey, the non-living philosopher that contemporary philosophers most identify with is Hume. See: https://philpapers.org/surveys/demographics.pl (accessed September 25, 2019).

Introduction

1   There are certainly many contributions concerning other prominent early modern figures' relations to natural philosophy and philosophy of physical science. To mention just a few: *Descartes' System of Natural Philosophy* by Stephen Gaukroger (2002), *John Locke and Natural Philosophy* by Peter R. Anstey (2012), and *Kant's Philosophy of Physical Science*, edited by Robert E. Butts (1986). However, there are no books like this on Hume.

2   Nearly all of Hume's works mention "natural philosophy," as the name appears in the *Treatise* and its Abstract, in both of the *Enquiries*, in the *Essays*, and in the *History*. It does not appear in the *Dialogues*.

3   As Katherine Brading (2015: 14) puts it, physics and philosophy had (and arguably still have) important "overlapping domains of investigation." For a thorough analysis of the status of science and philosophy in Newton, see Janiak (2015: chapter 2). Note that I do not claim that philosophy and science are the same thing. I think the two are different. What I object to is that there would be a dichotomy, that is, an all-encompassing distinction between them.

4   It should be noted, however, that Newton's work was instrumental to transforming the old natural philosophy into the modern specialized scientific discipline of physics (see Cohen, Smith, 2002: 1–4, and Grant 2007: 314–15).

Chapter 1

1   I will not discuss philosophy's relation to other natural sciences which would be relevant for natural philosophy, like chemistry and biology, as these disciplines fall outside the scope of this book.

2 Nicholas Maxwell's (2019) *Aeon* essay "Natural Philosophy Redux," and the academic article it is based on (Maxwell 2012), are similar in spirit to this chapter. Maxwell's description of natural philosophy as "a synthesis of physics of metaphysics, science and philosophy" is very close to the conclusions arrived at in this chapter. However, I do not agree with Maxwell that Newton killed natural philosophy while creating modern science. Even the third edition of the *Principia* includes a considerable amount of metaphysics (and theology). It is controversial to say that *Principia* diverged from natural philosophy altogether. I am also very sympathetic to Heather Dyke's (2007) views, as she presents them at her "Science and philosophy: Making time for each other."

3 In Snow's original phrasing, the two cultures are composed of "literary intellectuals" and "scientists." However, Snow did not support the divide but wished to allay it.

4 Defining science is certainly a difficult problem but here we are dealing with physics, which is something like an exemplar of science. We do not doubt whether physics is a science in the same way as we do not doubt whether the athletics of the Antiquity is sports. Activities like track, boxing, javelin, long jump, and the like are exemplars of sports. Of defining science by using exemplars and family resemblance in the spirit of Ludwig Wittgenstein's (1958) later philosophy, see Massimo Pigliucci (2013: 19–20).

5 For a recent defense of the heuristic function of philosophy in the development of scientific theories, see De Haro (2019, section 4).

6 Relatedly, Sam Baron (2018) has provided an interesting apology for metaphysics.

7 E. J. Lowe (2011: 99–100) defines his version of an aprioristic metaphysics as "a primarily *a priori* discipline concerned with revealing, through rational reflection and argument, the essences of entities, both actual and possible, with a view to articulating the fundamental structure of reality as a whole." Tim Maudlin (2007: 1) squares the naturalistic approach: "metaphysics, insofar as it is concerned with the natural world, can do no better than to reflect on physics. Physical theories provide us with the best handle we have on what there is, and the philosopher's proper task is the interpretation and elucidation of those theories." Note that the divide between the two approaches to metaphysics is by no means exhaustive. Matteo Morganti and Tuomas Tahko (2017) provide something like a middle position, which they call "moderately naturalistic metaphysics." For a complementary approach to the relation between science and metaphysics, see also Cláudia Ribeiro (2015).

8 We might of course consider figures who are included in the philosophical canon but whose works still contributed a great deal to physics, like Descartes, Leibniz, and du Châtelet. These philosophers did both theoretical and experimental research on conservation laws, optics, and dynamic concepts. As the three are primarily philosophers (for us, at least), their works are replete with metaphysics, logic, and epistemology. Instead of using these as examples of contributions to natural

philosophy, I shall instead consider Newton's input. He factors prominently in the history of physics and mathematics, but not philosophy. His work is primarily scientific, but there are a variety of philosophical elements in his thought; we can find a gray area in which philosophy and physics overlap. This is, in my understanding, the proper field for natural philosophy.

9   In the Axioms, or Laws of Motion, Newton's second law is, in our terminology, $\vec{F} = \Delta \vec{P}$. Here Newton treats the impressed force as an instantaneous, impulsive force. After approximating Kepler's Area Law with a polygon in Proposition 1 of the first Book, Newton writes that "the centripetal force by which the body is continually drawn back from the tangent of this curve will act uninterruptedly." Here he is clearly referring to the impression of a force in a given time, corresponding to our familiar $\vec{F} = \dfrac{d\vec{P}}{dt}$. Cohen (1999: 112) points out that in proposition 24 of the second book of the *Principia*, Newton writes that it is "manifest from the second law of motion" that "the velocity that a given force can generate in a given time in a given quantity of matter is as the force and the time directly and the matter inversely." See Cuicciardini (1998) for discussion.

10 He already equated the centripetal force and the force of gravity in Definition 5 of the first Book.

11 Newton tells us that a central objective of "philosophy seems to be to discover the forces of nature from the phenomena of motions and then to demonstrate the other phenomena from these forces" (ibid.: 382). The term "philosophy" is clearly not used in its present meaning in this context.

12 On the other hand, Newton's contribution to natural philosophy is in stark contrast with the Aristotelian tradition. Importantly, the argument for the law of universal gravitation falsified the Aristotelian distinction between lunar and sublunary worlds.

13 At the very end of the *Principia*, Newton mentions that maybe there is "a certain very subtle spirit pervading gross bodies and lying hidden in them," which would be causally responsible for all sorts of motions. But he adds that "there is not a sufficient number of experiments to determine and demonstrate accurately the laws governing the actions of this spirit."

14 Although some introductory expositions on the history of science (for example, Pine 1989; DeWitt 2010; Millican 2007) portray Newton as an instrumentalist concerning the force of gravity, recent scholarship on Newton (Janiak 2007; Kochiras 2011; Belkind 2012; Ducheyne 2012) emphasizes Newton's insistence on the reality of forces, on forces as true causes of motion. It is true that Newton is dissatisfied with action at a distance without a contact mechanism and that he does provisionally refer to forces and centers of masses only in mathematical rather than physical terms (*Principia*, first Book, Definition 8; see also Westfall 1993: 188). I. Bernard Cohen clarifies this issue as follows: In his research on the relation of centripetal forces to Kepler's area law, Newton begins his research by inquiring into mathematical rather

than physical proportions of forces. After this, he concludes that "a mathematically descriptive law of motion" is "equivalent to a set of causal conditions of forces and motions," as Cohen (1980: 28) puts it.

15 For a discussion on this point, see Janiak (2006: Section 1).

Chapter 2

1 Locke states the point explicitly in his *Essay* (2.1.2): "*All* ideas *come from sensation or reflection.*"

2 Whether the missing shade of blue is an exception is a somewhat controversial matter. Karánn Durland (1996) argues that successive color impressions might cause a complex idea in the mind. Complex ideas are not derived from simple impressions. This reading suggests that Hume does not allow in practice an exception to the copy principle (although he allows the theoretical possibility by means of fictitious ideas).

3 In Don Garrett's reading, Hume is pioneering cognitive psychology. Hume addresses the relation of imagination to memory, reason, and understanding. In the formulation of Garrett (1997: 39), this counts as a form of cognitive psychology because the acts of the understanding, including conception, judgement, and reasoning, are aspects of conceiving. These aspects make up the operations of representational faculty of imagination. This is an elementary theory of cognition. However, it is questionable whether this counts as a scientific discipline, because Hume did not carry out any actual empirical research. For example, he did not send out questionnaires (Garrett 1997: 48). There are also some phenomenological features in Hume's science of humanity. To quote from David Woodruff Smith's (2013: section 1) description of phenomenological study, "phenomenology studies the structure of various types of experience ranging from perception, thought, memory, imagination, emotion, desire, and volition to bodily awareness, embodied action, and social activity, including linguistic activity." The former description is not far away from Hume's announced objective. Nearly all the above-mentioned topics, perception, thought, memory, imagination, desire, volition and social activity figure prominently in the *Treatise*. However, I do not wish to fully equate Hume's science of human nature to the study of phenomenology. Experience in the phenomenological account involves what Husserl calls intentionality. Experience is always directed towards something in the world. Such intentionality is a property of consciousness, which is consciousness about something (Smith 2013: section 1). In Hume we do not find a phenomenological analysis of intentionality or consciousness.

4 As our perception (e.g., visual) is always complex (mixed with other sensations, and also spatial and temporal dimensions) we do not in fact perceive perfect simple impressions. David Landy (2018) argues that for Hume simple impressions are

explanatory posits or theoretical entities. Hume's science of humanity—together with a nominalist background metaphysics—explains the vast array of thoughts and experiences we have in terms of these simple impressions, although our perception is never so pure as to provide nothing but a simple, discrete, sensible impression.

5 Here "natural philosophy" should not be equated directly with "physics." Hume does not contribute to physics but to natural philosophy or philosophy of science.

6 Boyle occupies an interesting middle position between the experimentalist and speculative natural philosophers. He belongs to the experimentalist tradition as he supported experimentation and probable reasoning as ways of acquiring factual knowledge of nature, instead of the light of reason or axiomatic principles. Natural philosophy is restricted to matters of fact; it cannot reach demonstrative certainty via first principles (Shapin and Schaffer 1985: chapter 2). But he also made inferences to the best explanation in positing a microstructure of matter, the corpuscularian texture of bodies. Experimentalists like Hume later classified this as an imaginary pursuit (*History* V: 542).

7 The existing student lecture notes from the course indicate that the students saw experiments being carried out. Also, the University of Edinburgh had acquired a grant in the early eighteenth century from the Town Hall "for confirming and Illustrating by experiments the truths advanced in the Mathematicks and Naturall Philosophy" (Barfoot 1990: 153–4).

8 This is apparent in Hume's treatment of mixed mathematics in the first *Enquiry* (4.13; SBN 31–2), in which he uses a geometrical exposition of the law of conservation of momentum.

9 For a discussion of whether Hume wrote one of the papers and Wallace commented on it, see Gossman (1960).

10 Hume does use the term "law of nature" in the third Book of the *Treatise* in the context of political economy. Natural laws are then not solely confined within physics for Hume. Here I am referring to the causal relations among physical objects, not business transactions or such matters.

11 Here I am referring to Descartes' (1983) *Principles of Philosophy*, which establishes the similarity between space and body (see Chapter 6), hence excluding the vacuum, and to Leibniz's correspondence with Clarke, in which Leibniz repudiates Newtonian absolute space and time. In the view of Leibniz (1989b: 325), absolute space is a "chimerical supposition of the reality of space in itself." Likewise, he repudiates absolute time: "instants, considered without the things, are nothing at all [...] they consist only in the successive order of things."

12 This is also Schliesser's (2009) position. Hazony (2014) argues that for Hume metaphysics is psychology, and this psychology is the foundation for all other sciences. Hazony (ibid.: 145) compares Hume's system of the sciences to Descartes' tree of philosophy.

13 Traiger's (2018) position is also relevant in this context.

14 Although, if substance is defined as something that exists by itself, only God is a substance in Descartes.

15 At least Hume does not explicitly use the words "analytic" and "synthetic." In recent scholarship, Millican (2017: 3, fn. 2) argues that Hume's relations of ideas are equivalent to analytic propositions, as both can be thought as being true by the meanings of the terms involved. For their part, synthetic propositions are not true solely by meaning of the component terms.

16 After Hume had presented his dichotomous distinction of relations of ideas and matters of fact in the fourth section of the first *Enquiry*.

17 This is the traditional reading of Hume's metaphysical position. For his background in the nominalist tradition, see Deborah Brown (2012). For an alternative, trope theoretical reading, see Jani Hakkarainen (2012a).

18 Note that this is not Baxter's view. He argues that Hume is both a Pyrrhonian skeptic and a constructive metaphysical theorizer.

Chapter 3

1 The research project is *Early Modern Experimental Philosophy. A Project of the Early Modern Thought Research Theme at the University of Otago.* The team members include Alberto Vanzo, Juan Gomez, Kirsten Walsh, and Peter Anstey. The project has a blog: https://blogs.otago.ac.nz/emxphi/

2 In his listing, Vanzo leans on the following scholarship: Stephen Gaukroger (2010: 156), Knud Haakonssen (2004: 102, 109–14), Desmond M. Clarke (1982), Harry M. Bracken (1974: 15–17, 259), and Stephen H. Daniel (2007: 163–80).

3 This was already argued by Mary Shaw Kuypers ([1930]1966: v) in 1930. In the preface to her seminal work, she notes that standard expositions of "Hume's philosophy have treated it chiefly in its relation to the English epistemological tradition." She notes that at the time no detailed studies of Hume's connection to early modern science had been done.

4 Hume does not use the notion of tautology. To my knowledge, Wittgenstein (1922) was the first to introduce this notion in his philosophy of logic and mathematics in his *Tractatus* (especially under the sections 4, 5, and 6). Along with contradiction, tautology is the other extreme truth condition of propositions. Tautologies do not have truth conditions because they are unconditionally true. A tautologous proposition cannot picture reality because it allows all possible states of affairs (*Tractatus* 4.461–2). The proposition "It either rains or it does not" does not convey any information about weather conditions—or indeed about anything factual. Although Hume does not use the word tautology, his philosophy of mathematics

and overall epistemology express a similar idea to Wittgenstein. For Hume mathematical propositions are made of clear and distinct ideas that form a unity. This is rather close to the notion of a tautology, since such unitary ideas are devoid of all factual content; they do not refer outside themselves.

5   Although vortex cosmologies are difficult to square with Kepler's elliptical laws. See Smeenk (forthcoming).

6   Leibniz's mechanistic physics is not in contradiction with Newton's reported experiment of his third law in the Scholium to the Axioms, or Laws of Motion in the *Principia*. Leibniz does not find evidence that the third law could be extrapolated from tested instances of colliding or contiguous objects to celestial ones which are separated by a distance.

7   It should be emphasized that intelligibility is a context sensitive matter, see de Regt (2017: 40). Here the relevant context is the mechanistic natural philosophy.

8   Vortex experiments might not be entirely impossible. Descartes' astrophysics is utterly speculative but Huygens arranged an experiment to explain gravity with vortexes. The experiment used a vessel filled with water (modelling ether) and pieces of sealing wax (modelling planets) in the water. Rotating the vessel should imitate the real astronomical vortex-like motions (Snelders 1989: 212–13).

9   I am not arguing that Newton was uninterested in finding the reason for gravity, or that there is not, even in principle, any reason for gravity's operations (see Sapadin 2009: 80). Moreover, Newton denies that gravity is an occult quality. Gravity's cause may be found out. In his preface to the *Principia*'s (1999: 392) second edition in 1713, the editor Roger Cotes supports Newton on this point: "I can hear some people disagreeing with this conclusion and muttering something or other about occult qualities. They are always prattling on and on to the effect that gravity is something occult, and that occult causes are to be banished completely from philosophy. But it is easy to answer them: occult causes are not those causes whose existence is very clearly demonstrated by observations, but only those whose existence is occult, imagined, and not yet proved. Therefore gravity is not an occult cause of celestial motions, since it has been shown from phenomena that this force really exists. Rather, occult causes are the refuge of those who assign the governing of these motions to some sort of vortices of a certain matter utterly fictitious and completely imperceptible to the senses."

10 Both Descartes and Leibniz advance such speculative astrophysics. In his *Principles*, Descartes imagines comets wandering between different stars being strapped by large circling bands, the vortexes. In his *Tentamen de Motuum Celestium Causis*, Leibniz devised a vortex-based cosmology, which also encompasses Kepler's area law and Newton's centripetal force. Both of these hypotheses are non-empirical, but alluring because they render the phenomena intelligible.

11 Importantly, Miller argues that not only are the qualities of bodies, like impenetrability, hardness, and inertia, ascribable to all bodies universally by means

of induction, but that Newton's Rule 3 generalizes the relations between bodies as universal laws. This enables Miller to explain how the force of gravity is not a quality of an individual body, but a relation among bodies. At the very end of his Rule 3, Newton makes the point that gravity is not essential to bodies. This suggests, as Miller points out, that the inductive reasoning establishes the universality of the law of gravitation, not that gravity is a property of each body. On the Hume–Newton connection regarding induction, see also Belkind (2019).

12 For a list of Hume's references to Newton, see Force (1987: 169–77).

13 Here Hume's proud acknowledgement of his ignorance is consistent with the modesty of the experimental narrative. See Shapin and Schaffer (1985: 65).

14 Thus Newton puts it as follows in the General Scholium: "This most elegant system of the sun, planets, and comets could not have arisen without the design and dominion of an intelligent and powerful being." In Newton's view, motions of objects are governed by a causal chain. He is explicit about the fact that gravitational "motions do not have their origin in mechanical causes." The proximate cause for a motion of an object is an impressed force or gravity. Gravity has a proximate cause. The ultimate, most remote cause for motions of objects is God (Ducheyne 2012: 22).

15 However, analogical reasoning assumes the uniformity principle.

16 The negations of relations of ideas are inconceivable, whereas the negations of facts are conceivable. As I will argue in Chapter 5, which deals with Hume's notion of mixed mathematics, another distinguishing feature is that relations of ideas are not dependent on the uniformity principle, whereas facts do depend upon it.

17 For uses of Adam in the medieval tradition, see Rosier-Catach (2016), and in the early modern period, see Harrison (2002).

18 Like Boyle, Hume (EHU 10.4; SBN 110–11) also finds the number of experiments to be relevant for the establishment of facts: "A hundred instances or experiments on one side, and fifty on another, afford a doubtful expectation of any event; though a hundred uniform experiments, with only one that is contradictory, reasonably beget a pretty strong degree of assurance."

19 Owen (2002: 339) assumes that Hume's maxim in his critique of reported miracles— no testimony may establish a miracle, unless its falsehood is more miraculous than the putative miracle it tries to establish (EHU 10.13; SBN 115–16)—conforms to the following version of Bayes' theorem: $\dfrac{pt}{pt+(1-p)(1-t)}$, in which p means the prior probability of the event that is being testified, and t is the probability of the truth-likeness of the testimony.

20 Such conditional probability could be taken, for example, from the tunneling effect in quantum mechanics. The probability of even a proton tunneling a microscopic distance is significantly smaller than 10^{-1000}. An object suspended in the air will fall down with a staggeringly high probability. In my example, the conditional probability for a very good witness getting her testimony right was estimated very

highly. My objective here is not to provide an extensive explanation of Hume's use of Bayesian inference (this is done, for example, by Owen 2002). Rather, I wish to illustrate the connection between experience, the social-testimonial nature of evidence, and subsequently different range of probabilities that may be assessed to propositions concerning matters of fact. Here Hume's views are clearly related to Boyle's experimentalism, which is what this chapter aims to show: Hume's epistemology is deeply embedded in the previous tradition of experimentalist natural philosophy.

Chapter 4

1 This point can be extended to Descartes' astrophysics. It was not quantitative, because Descartes did not assume that the solar system(s) remain steady in the course of long periods of time (Smeenk, forthcoming).

2 For a list of these positions, see chapter 7, "Laws and causation."

3 Again, it is useful to refer to Hume's thought example on Adam. In David Fate Norton's (2006: 29–33) commentary, Adam is following a game of pool. The game begins as the starting party shoots the cue ball toward the pack. What does Adam expect to happen in the upcoming collision? Nothing. Not one little thing. He has never seen moving billiard balls. What he sees now is merely a moving, colored blob. It is beyond Adam's knowledge that this blob hits a bunch of similar blobs and initiates their motion. Norton suggests that if the table were sufficiently long, and we were to ask Adam what happens after the collision, Adam would respond: "What is a collision?"

   Hume's answer as to why we anticipate the object balls to move after impact is that we have frequent experience. By this we form a habit of inference. Adam could reason the same way, if he were to observe the constant conjunction of the balls repeatedly. Mere contiguity and succession of the balls does not establish a causal relation. Hume writes that they "are not sufficient to make us pronounce any two objects to be cause and effect, unless we perceive, that these two relations are preserv'd in several instances" (T 1.3.6.3, SBN 87–8). He should witness the following sequences, in which C denotes the cue ball, and O an object ball (Norton 2006: 31):

   C_1 collides to O_1, and then O_1 initiates a motion;
   C_2 collides to O_2, and then O_2 initiates a motion;
   C_3 collides to O_3, and then O_3 initiates a motion;
   C_n collides to O_n, and then O_n initiates a motion . . .

   For a thorough treatment of Adam's role in Hume, see also Robison (2018).

4 It should be noted that Boehm (2018) does not identify with projectivism or the old or new Hume readings. Instead, she supports the foundational project interpretation.

In her rendition, Hume does not argue for a metaphysical theory of causation but examines causation from the viewpoint of his science of human nature.

5 In the Abstract (25; SBN 655–6) to the *Treatise*, Hume makes a similar point about discovery of causation in natural philosophy: "We have confin'd ourselves in this whole reasoning to the relation of cause and effect, as discovered in the motions and operations of matter."

6 These are all dynamic laws so the list does not include references to any optical laws.

7 Chambers *Encyclopedia* (1728: 521) has a very similar formulation of the general conservation law (as expressed in Hume's rendition of definition in law three): "Wherefore in any Machine or Engine, if the Velocity of the Power be made to the Velocity of the Weight: reciprocally as the Weight is to the Power; then shall the Power always sustain, or if the Power be a little increas'd, move the Weight."

8 This issue is closely related to the debate of whether Newton used the notion of a force field. Stein (1970) is a classical article defending the view that Newton did use the concept of a gravity field. Schliesser (2011) challenges Stein's reading. Schliesser argues that Newton understood forces as interactions, so a single particle alone cannot create a gravity field.

9 In Definition 10 of his "De Gravitatione," Newton claims that "gravity is the force in a body impelling it to descend." This suggests that there has been a change in Newton's thought: before the *Principia* he saw force to be something in a body, whereas in the *Principia* force is an interactive relation among masses. In Rule 3 of the Rules for the Study of Natural Philosophy in the *Principia*, Newton explicitly says that gravity is not essential to bodies.

10 If we take into account the theological context, then Newton would not argue for the total emptiness of space. This is because he believes in God's omnipresence (Newton 1999: 940–1).

11 Here Malebranche's discussion on causation in his *Search after Truth* is a probable source of influence on Hume.

12 On this issue, Descartes' and Hume's argumentative paths are certainly very different. Descartes' argument is a corollary of his metaphysics of substance. There are two substances in the world: minds and bodies. The attribute of a mind is thought, whereas the attribute of a body is extension. Hume's argumentation is based on his copy principle. The mind has a threshold in forming adequate representing ideas. There are minimum sensible items, extended finite tactual or visual points, which are the source for the (abstract) idea of space. Although the two authors have markedly different ways in which they reach their conclusions, they nevertheless both conclude that space is extension. Assimilating space and extension makes the notion of a vacuum, or empty space, questionable.

13 Whether Hume completely rejects action at a distance is not obvious. In *Treatise* 1.2.5.26n (SBN 638–9), he remarks that it is not clear "whether or not the invisible

and intangible distance be always full of body." He does not find a "very decisive argument on either side." (For a thorough reading on fictitious distance, see Boehm (2012)). There can be causally related existent objects even though the objects are not spatial beings, as Hume's metaphysical maxim in T 1.4.5.10 (SBN 235), "*an object may exist, and yet be no where*" states. However, when it comes to bodies, Hume's talk of an object being "ever so little remov'd" (T 1.3.2.6; SBN 75) seems to in fact indicate that for bodies to be causally related there must be contact among them. Two bodies cannot be considered as causes and effects if they do not touch each other. Hume is explicit on this matter in the Abstract 9 of the *Treatise*, where he says that "contiguity in time and place is therefore a requisite circumstance to the operation of all causes" (SBN 649). He highlights the need of physical contact, that is, touch, when he addresses the causal relation among two billiard balls: "The first ball is in motion; touches the second; immediately the second is in motion: and when I try the experiment with the same or like balls, in the same or like circumstances, I find, that upon the motion and touch of the one ball, motion always follows in the other" (Abstract 9; SBN 650). Causation regarding bodies is of course different from events, which only need to be in next-to relation to each other (for this point, see Slavov 2016c).

14 Except that magnetic attraction is a non-mechanical force, which Hume does tacitly accept (EHU 4.7; SBN 28). Magnetic force is a long-range force like gravity, but intuitively it might be easier to accept that a magnet attracts pieces of metal in its vicinity, than accepting that the Sun moves Jupiter.

15 Or at least we should have perceived either the cause or the effect in the past. If an observer were to go to a deserted island, she would be able to infer that a footprint on the sand was left there by a human being: "If you saw upon the sea-shore the print of one human foot, you would conclude, that a man had passed that way" (EHU 11.24; SBN 142–3). This is because the observer has previous experience of how the relevant objects, the foot and the sand, interact.

16 In his "An Essay Towards a New Theory of Vision" (section 85), Berkeley (1948) argued for the relativity of perception to current microscopic technology: "For when we look through a microscope we neither see more visible points, nor are the collateral points more distinct then when we look with the naked eye at objects placed in a due distance. A microscope brings us, as it were, into a new world: it presents us with a new scene of visible objects quite different from what we behold with the naked eye."

17 As Jesseph (1992: fn. 11) notes, Berkeley is here presumably referring to *Principia's* definition 8, where Newton writes about forces: "This concept is purely mathematical, for I am not now considering the physical causes and sites of forces." I find it possible that Berkeley could also be referring to Section 11 of the first book where Newton treats attraction of two bodies toward a "common center of gravity"

in the following way: "For here we are concerned with mathematics; and therefore, putting aside any debates concerning physics, we are using familiar language so as to be more easily understood by mathematical readers."

18 Another way to interpret Berkeley's conclusions is to say that his objective is to reduce dynamics to kinematics. See Downing (2005: 263–4, fn. 51).

19 For this reason, as Downing (2005: 253) points out, Berkeley preferred "Newtonian physics to Cartesian physics."

20 Mary Shaw Kuypers ([1930]1966: 79) argues that, as forces are not observable, the inferences including a reference to forces assume the principle of necessary connection: "If force is an existent quality or entity resident in objects, it should be discoverable through sensation; for if it exists beyond experience, it can be inferred only by assuming the principle of necessary connection, which was ruled out by Hume's argument against the logical necessity of cause." I disagree with the latter part of Kuypers' reading. Regular past experiences give us a basis to assess conditional probabilities of factual propositions; this does not require that facts are necessary.

Chapter 5

1 On the fluctuation of Hume's views on geometry, see Batitsky (1998).

2 As Baxter (2007: 2) formulates the problem of identity, many different things should be one: "Hume says that thinking of things *a* and *b* as identical requires being able to think of them as one yet also as many, and so requires thinking inconsistently." Hume's position is that a complex is not in reality a single thing but many things (ibid.: 17).

3 It may be noted that Hume allows that there is one exception where a proposition concerns only one idea. In a footnote to the first Book of the *Treatise*, he claims that propositions regarding existence can be formulated only by one idea. In propositions such as "an object x exists," "the idea of existence is no distinct idea, which we unite with that of the object." Thus, we can "form a proposition, which contains only one idea" (T 1.3.7.5; SBN 96f., fn. 20). Baxter (2007: 57) indicates that these are mere "trifling" propositions. To say that an object is the same with itself does not bring forth any new information concerning any fact. Thus Baxter puts it as follows: "In general, a trifling proposition is one in which the proposition as a whole adds nothing to the idea that is the subject." This is different from learning about the qualities of objects. I know that, for example, the coffee I poured from the pan to my mug is hot because I have experience of the relevant causal relation, "coffee causes the sensation of hotness." The hotness of coffee is not self-evident, trivial inference, because such black liquid may have an indefinite amount of properties that can be predicated of it.

4 Because Hume bases his philosophy of mathematics on the early modern theory of ideas, it may be that he did not have the substitutional view of logic. According to the definition of Adriane Rini (2016: 12–13), "on a substitutional view of logic, an inferential schema is valid if and only if every possible substitution instance of it is truth-preserving. When we put terms into the valid schemas and generate premises involving actual truths, then the conclusion itself is always going to be an actual truth." For Hume, every term in a syllogism would be an idea (although linguistic terms are different from ideas, because language and thought are distinct for Hume). We reason from an idea to its adjacent idea. In this view, there is no abstract logical structure that maintains the validity of logical inferences. There is no logical structure over and above the ideas.

5 In the *Treatise*, Hume does not explicitly claim that the denial of mathematical propositions is contradictory. See Steiner (1987: 402).

6 In the view of Millican (2017: 29–30), by the criterion of demonstrability of relations of ideas, there are no false mathematical propositions.

7 Regarding matters of fact propositions in the first *Enquiry*, Hume focuses on causation. The first *Enquiry* does not deal with relations of identity, and space and time, as in the first Book of the *Treatise*.

8 Whether Popper's argument evades induction is unclear, as the notion of corroboration seems to incorporate inductive logic. Popper might smuggle induction into his falsificationism. See Morris (2011: 460).

9 It should be added that laws are causal, and that identifying causation requires more than just tracking regularities. As Hume points out in his *Treatise* (1.3.15.2; SBN 173): "Since therefore 'tis possible for all objects to become causes or effects to each other, it may be proper to fix some general rules, by which we may know when they really are so." So, in addition to tracking regularities, the identification of laws of nature also requires rules which distinguish mere regularities from causal action. For example, "the night is followed by a day" is a true regularity about nature, but it is not a causal relation, or constant conjunction. Instead, such regularity instantiates two phases like two sides of a coin.

10 Regarding the meaning of "proof" in Hume, I follow De Pierris' (2006) interpretation.

11 Newton's second law and gravity law are written here in presentist terms. In Newton's second law, the impulsive force is proportional to the change of momentum. The single formula for universal gravity is not in the *Principia*. For a discussion on notation, see Cushing (1998: 98–9, 108–9). Still it is reasonable to assume that Hume refers to Newton's second law and gravity law, as he says that "the production of motion by impulse and gravity is an universal law" (EHU 6.4; SBN 57).

12 Carla Rita Palmerino (2016: 30–1) goes through interpretations which deny that Galileo would have consistently held the mathematical structure of reality. Then she (2016: 32) goes on to argue that "in his works Galileo repeatedly argues that

mathematical entities are ontologically independent from us and that the physical world has a mathematical structure."

Chapter 6

1　Rynasiewicz (2014) points out that for Newton space and time are not substances in the paradigmatic sense, like bodies or minds, but their self-sustaining existence is necessitated by God's omnipresence and eternality. For Newton's rejection of Cartesian dualism, see Janiak (2013: Section 3). As absolute space and time are not substances, I use the term "absolutist argument," instead of the common notion of "substantivalism about space and time."

2　This is Schliesser's (2013) argument.

3　For a classical defense of metric conventionalism, see Grünbaum (1973). For a critical evaluation of metric conventionalism about time, see Dowden (2018: Section 25).

4　For the theological underpinnings of Newton's argument, see Schliesser (2013: 95).

5　This is commonsensical because it is *unlike* the mind-boggling ramifications of the relativity of simultaneity in the theory of special relativity. The relativity of simultaneity of space-like related events leads (or many have argued) to the astonishing view that the distinction between the past, present, and future is subjective, dependent on the choice of the frame of reference. According to the eternalist interpretation, it is not that the past is gone, the now exists, and the future is yet to become. Instead, past, present, and future are all equally real.

6　See Mikkeli (2018: Section 5) for the *regressus*-method.

7　Eventually, the kinematical case is also *not* perspectival for Newton. He is utilizing the notion of absolute velocity. There is a categorical difference between: *rest*—no motion (zero velocity) with respect to absolute space; *inertial motion*—uniform motion (constant velocity) with respect to absolute space; *acceleration*—non-uniform motion (changing velocity) with respect to absolute space.

8　Technically, God is the only substance in Descartes' metaphysics and theology, because He exists independently of anything else.

9　Descartes writes: "the extension in length, breadth, and depth which constitutes the space occupied by a body, is exactly the same as that which constitutes the body" (Pr II 10).

10　Descartes makes a distinction between "place" and "space" in a sense that the first denotes the location of the body, and the latter the attributes of extension, like size and shape (Pr II 14). However, this certainly is in tension with Newton's definition of place, according to which bodies fill empty space. Descartes clarifies his position in stating that a place cannot be absolutely nothing. An empty jar is not absolutely empty, as it contains air; a fish-pond without fish is not absolutely empty, as it contains water; and a cargo ship is not absolutely empty, although it does not carry any merchandise (Pr II 17).

11 For the logically valid argument from limited capacity, see Cottrell (2018: 87–8).

12 In Hume's own example (T 1.2.1.4; SBN 27–8): "Put a spot of ink upon paper, fix your eye upon that spot, and retire to such a distance, that at last you lose sight of it; 'tis plain, that the moment before it vanish'd the image or impression was perfectly indivisible." As noted in Chapter 4, Hume does not beforehand stipulate the minimum size of the being that is represented. Microscopic and telescopic technologies enlarge our sensory perception: "A microscope or telescope, which renders them visible, produces not any new rays of light, but only spreads those, which always flow'd from them; and by that means both gives parts to impressions, which to the naked eye appear simple and uncompounded, and advances to a *minimum*, what was formerly imperceptible" (ibid.).

13 In Paul Russell's (1997) seminal reading, Hume's target is Clarke's Newtonian creed of absolute space and his *a priori* argument for the existence of God. So, in this respect, Hume's critique of Newtonianism has important theological dimensions. Cottrell (2018: Section 1) also analyzes the importance of the theological context.

14 Visual and tactile impressions are discrete in principle, but in practice our perception of extension is a mix of sensations of vision and touch. Landy (2018: 34) notes that copy principle is an idealized principle in this sense: our perception is not in fact about discrete simple impressions, but Hume uses simple perceptions as "posits" or "theoretical entities" to explain perception.

15 Hakkarainen adds that this is along the lines of Hume's general position on the metaphysics of perceptions. In the *Treatise* (1.4.2.47; SBN 212), Hume proclaims that "the only existences, of which we are certain, are perceptions, which being immediately present to us by consciousness ... no beings are ever present to the mind but perceptions."

16 Concerning absolute and metaphysical modalities in T 1.3.14.35 (SBN 172), see Holden (2014) for an in-depth analysis.

17 For clarification, see Baxter's (2007: 37) brick wall diagram.

18 By this assimilation I am not suggesting that Hume is an anti-realist about time like McTaggart.

19 This case is analogous to our belief in synchronic personal identity. There is no sensory evidence, for example, for a permanent substance of the mind in which perception inhere. Still we in our everyday lives subscribe to "an owner view" of our mental states. For a comprehensive list of fictions of the imagination, see Cottrell (2018: section 5). Moreover, Cottrell (2015: 94) argues that the illusion we have on absolute space is analogous to absolute time: time without change, or an object enduring without a change, is a "temporal vacuum" for Hume.

20 For the affinity of Descartes' ontology with relativity, see Slowik (2005).

Chapter 7

1 The group gathered around 1902–1903 in Bern. In addition to Einstein, it included philosophy student Maurice Solovine and mathematician Conrad Habicht (Howard 2005: 36; Janssen and Lehner 2014: 2).

2 Later, Einstein (2002: 20) went so far as to claim that "the phenomenon of magnetoelectric induction compelled me to postulate the (special) principle of relativity." See also Einstein (1981a: 218; 1981h: 320–1).

3 See Galison (2003: chapter 3).

4 The examples Kuhn (1996: 38) uses of the sciences preceded and accompanied by philosophical analyses include Newtonian physics, relativity, and quantum mechanics.

5 Einstein had been devising thought experiments about electrodynamic phenomena, which include a reference to the ether, as early as 1895. See Einstein (1987: 4–6), "On the Investigation of the State of the Ether in Magnetic Field."

6 Einstein's term "concept" (Begriff) and Hume's term "idea" cannot be used interchangeably: "concept" is more definitional and conventional, whereas "idea" is given to the mind, and our will regarding it is not free. However, Einstein sometimes conflates the two terms. Although he usually speaks about "concepts" (see Einstein 1981b: 221; 1981d: 241; 1981f: 270; 1981g: 284, 286; 1949a: 13; 2001: 24), in his *Meaning of Relativity* he begins with "investigation of the origin of our *ideas* of space and time" (Einstein 2003: 1, my emphasis). Further, Hume's term "impression" is more extensive than Einstein's "sensation": Hume's "impression" covers feelings and emotions, not just "sensations," if by sensations is meant "comes from the senses."

7 In this context, the empiricist argument is coupled with a conventionalist argument concerning the one-way speed of light. My work focuses on the empiricist argument; for a thorough treatment of conventional aspects, see Øhrstrøm (1980) and Einstein's (1923: 39–40) own argument in the original publication of STR.

8 One could argue that this does not yet debunk Newton's distinction between absolute and relative time, because he did not say that clock time is independent of relative motion but that time itself is independent of any change. Still, the Lorentz transformations include the relativity of simultaneity—if in one frame two events are simultaneous, they are not simultaneous in another frame—as the temporal interval between two events is not absolute.

9 I do not wish to state that the absolutist or transcendental arguments are blatantly wrong and obsolete. That would be a mistake. There are still good reasons to hold on to tempered forms of these positions. Absolute space and absolute time, understood separately, do not exist. But a more fundamental structure, spacetime, can be argued to exist. Although it is an open question as to what sense it can be said to exist, many physicists and philosophers of physics approve of an absolutist argument modified

by Minkowski's four-dimensional spacetime (Greene 2004: 61; for different formulations of the four-dimensional spacetime, see Gilmore et al.: 2016). In addition, the Kantian philosophy of time could be seen as consistent with eternalism, which is usually thought to be a cogent philosophy of time for STR. Four-dimensional spacetime does not make an objective difference between past, present, and future. The flow of time might be a subjective feature of our minds, and reality in itself might be a static block universe. This thought is plausibly consistent with Kant's transcendental idealism, which provides a distinction between the temporal phenomenal world and the atemporal noumenal world (Weinert 2005).

10 Similarly, in his "The Problem of Space, Ether, and the Field in Physics," Einstein (1981f: 271) argues that the concept of a body is not preceded by the notion of space: "Once the concept of the solid body is formed in connection with the experiences just mentioned [sight and touch]—which concept by no means presupposes that of space or spatial relation." He indicates that he has "never been able to understand the quest of the *a priori* in the Kantian sense" (ibid.). Einstein differs from Kant in that he does not think that space (and time) are *a priori* forms of sensibilities which precede all possible experience, such as the experience of bodies. Rather, Einstein clearly contends that a material object precedes the concept of space (and time) (Einstein 1981i: 355; see also Lenzen 1949: 367).

11 See also Hentschel (1992: 619–21).

12 This conclusion assumes that the one-way speed of light is constant. I do not take a stand on whether this is a convention or not.

13 By adopting Minkowski light cones, it can be said that, in the ship's frame, the castaway's sending of the message is in the absolute past, and in the castaway's frame, the ship's receiving of the information is in the absolute future.

14 In Hakkarainen's (2012b) reading, there is no one Hume on this issue. Rather, Hume-the-philosopher is skeptic of metaphysical realism, but Hume-the-everyday-person is a metaphysical realist. Hume suspends his judgment on philosophers' realism, but in common life he believes,, without a doubt, in mind-independent entities.

15 Markus Schrenk introduces the main tenets of the two positions and lists their main proponents in Philpapers' General Philosophy of Science section, "Humeanism and Nonhumeanism about Laws": http://philpapers.org/browse/humeanism-andnonhumeanism-about-laws. Accessed on December 12, 2018.

16 Dispositionalism, for example, Anjum's and Mumford's (2011) book *Getting Causes from Powers*, also counts as a form of NonHumeanism.

17 The laws of nuclear physics that Carroll refers to are also probabilistic.

18 Beebee advances this argument in a debate, "The Laws of the Universe": https://iai. tv/video/the-laws-of-the-universe. Accessed on September 24, 2019.

19 Hume of course drops the contiguity requirement in the first *Enquiry*. However, the first *Enquiry* still includes preconditions 2 and 3. See Schliesser (2007: Section 4.4).

20 As I noted earlier in Chapter 4, Newton's third law is not meant to be a causal law, but rather a law of coexistence (see Tooley, 2004: 87–8). Still, it is noteworthy to analyze it as it sheds light on the trouble of interpreting laws in causal terms.

21 It could be further objected that distinguishing between a cause and an effect in such a scenario is somewhat arbitrary. It is not clear whether cause and effect stands for objects or events, and where exactly does the cause end and the effect begin?

22 In the first *Enquiry* (7.29. SBN 76–7), Hume writes "we may define a cause to be *an object, followed by another, and where all the objects, similar to the first, are followed by objects similar to the second. Or in other words, where, if the first object had not been, the second never had existed.*" For the incompatibility of the regularity and counterfactual conditions, see Menzies (2014: Section 1).

23 For a Humean interventionist theory of causation, see Kuorikoski (2014).

24 For an overview of causation in physics, see Blanchard (2016).

25 On this point, see Slavov (2020).

26 Hume argues that simultaneous causation is a reduction to absurdity, not that it is impossible. For a thorough treatment of succession in causation, see Ryan (2003).

27 The calculation here is simplistic because the other observer accelerates. For an explanation of the so-called "Twins paradox," see Maudlin (2012: 77–83).

Bibliography

Anjum, R. L. and S. Mumford (2011), *Getting Causes from Powers*, New York: Oxford University Press.

Anstey, P. R. (2012), *John Locke and Natural Philosophy*, Oxford: Oxford University Press.

Anstey, P. R. and A. Vanzo (2012), "The Origins of Early Modern Experimental Philosophy," *Intellectual History Review*, 22 (4): 499–518.

Anstey, P. R. and A. Vanzo (2016), "Early Modern Experimental Philosophy," in J. Sytsma and W. Buckwalter (eds), *A Companion to Experimental Philosophy*, 87–102, West Sussex: Wiley-Blackwell.

Archimedes (1879), *On the Equilibrium of Planes*, trans. T. L. Heath, in *The Works of Archimedes*, 189–220, London: C. J. Clay and Sons.

Armstrong, D. M. (1983), *What Is a Law of Nature?* Cambridge: Cambridge University Press.

Armstrong, D. M. (2004), *Truth and Truthmakers*. New York: Cambridge University Press.

Ayer, A. J. (2001), *Language, Truth and Logic*, London: Penguin.

Bacon, F. ([1605]1808), *Of the Proficience and Advancement of Learning*, London: Parker, Son, and Bourn.

Barbour, J. B. (2007), "Einstein and Mach's Principle," in J. Renn (ed), *The Genesis of General Relativity*, 1492–527, Dordrecht: Springer.

Bardon, A. (2013), *A Brief History of the Philosophy of Time*, New York: Oxford University Press.

Barfoot, M. (1990), "Hume and the Culture of Science in the Early Eighteenth Century," in M. A. Stewart (ed.), *Studies in the Philosophy of the Scottish Enlightenment*, 151–90, Oxford: Clarendon Press.

Baron, S. (2018), "A Formal Apology for Metaphysics," *Ergo*, 5 (39): 1030–60.

Batitsky, V. (1998), "From Inexactness to Certainty: The Change in Hume's Conception of Geometry," *Journal for General Philosophy of Science*, 29 (1): 1–20.

Bardon, A. (2007), "Empiricism, Time-Awareness, and Hume's Manners of Disposition," *Journal of Scottish Philosophy*, 5 (1): 47–63.

Baxter, D. (2007), *Hume's Difficulty: Time and Identity in the Treatise*, New York: Routledge.

Baxter, D. (2015), "Descartes and Hume on Duration," *Proceedings of the 42nd International Hume Society Conference Stockholm*: 203–16.

Baxter, D. (2016), "Hume on Space and Time," in P. Russell (ed.), *The Oxford Handbook of Hume*, 173–90, New York: Oxford University Press.

Beebee, H. (2000), "The Non-Governing Conception of Laws of Nature," *Philosophy and Phenomenological Research*, 61 (3): 571–94.

Beebee, H. (2006), *Hume on Causation*, London and New York: Routledge.

Belkind, O. (2012), "Newton's Scientific Method and the Law of Universal Gravitation," in A. Janiak and E. Schliesser (eds), *Interpreting Newton: Critical Essays*, 138–68, New York: Cambridge University Press.

Belkind, O. (2019), "In Defense of Newtonian Induction: Hume's Problem of Induction and the Universalization of Primary Qualities," *European Journal for Philosophy of Science*, 9 (1): 1–26.

Beller, M. (2000), "Kant's Impact on Einstein's Thought," in D. Howard and J. Stachel (eds), *Einstein: The Formative Years, 1879–1909*, 83–106, Boston, Basel, Berlin: Birkhauser.

Berkeley, G. (1948), "An Essay Towards a New Theory of Vision," in A. A. Luce and T. E. Jessop (eds), *The Works of George Berkeley, the Bishop of Cloyne, Volume 1*, 171–236, London: Thomas Nelson and Sons.

Berkeley, G. (1992), "De Motu," in D. M. Jesseph (ed., trans), *De Motu and The Analyst*, 73–107, Dordrecht: Kluwer Academic Publishers.

Berryman, S. (2016), "Democritus," in E. N. Zalta (ed.), *The Stanford Encyclopedia of Philosophy*, https://plato.stanford.edu/archives/win2016/entries/democritus/

Blanchard, T. (2016), "Physics and Causation," *Philosophy Compass*, 11: 256–66.

Boehm, M. (2012), "Filling the Gaps in Hume's Vacuums," *Hume Studies*, 38 (1): 79–99.

Boehm, M. (2013a), "Hume's Foundational Project in the *Treatise*," *European Journal of Philosophy*, 24 (1): 55–77.

Boehm, M. (2013b), "The Concept of Body in Hume's *Treatise*," *Protososiology*, 30: 206–20.

Boehm, M. (2016), "Hume and Newton's Empiricism and Conception of Science," paper presented at the *43rd International Hume Society Conference*, University of Sydney.

Boehm, M. (2018), "Causality and Hume's Foundational Project," in A. Coventry and A. Sager (eds), *The Humean Mind*, 110–23, London and New York: Routledge.

Boyle, D. (2012), "The Ways of the Wise: Hume's Rules of Causal Reasoning," *Hume Studies*, 38 (2): 157–82.

Boyle, R. ([1660]1999), *New Experiments Physico-Mechanical, Touching the Spring of the Air and its Effects*, in M. Hunter and E. B. Davis (eds), *The Works of Robert Boyle*, Volume I, 141–302, London: Pickering and Chatto.

Boyle, R. (1675), "Some Considerations about the Reconcileableness of Reason and Religion," in T. Birch (ed.), *The Works of the Honourable Robert Boyle*, 151–91, London: J. & F. Rivington.

Boyle, R. (1685), *A Free Inquiry into the Vulgarly Received Notion of Nature*, London: H. Clark for John Taylor.

Boyle, R. (1989), "The Excellency and Grounds of the Corpuscular or Mechanical Philosophy," in M. R. Matthews (ed.), *The Scientific Background to Modern Philosophy: Selected Readings*, 111–23, Indianapolis/Cambridge: Hackett Publishing Company.

Brading, K. (2012), "Newton's Law-constitutive Approach to Bodies: A Response to Descartes," in A. Janiak and E. Schliesser (eds), *Interpreting Newton: Critical Essays*, 13–32, New York: Cambridge University Press.

Brading, K. (2015), "Physically Locating the Present: A Case of Reading Physics as a Contribution to Philosophy," *Studies in History and Philosophy of Science Part A*, 50: 13–19.

Brown, D. (2012), "Hume and the Nominalist Tradition," *Canadian Journal of Philosophy*, 42 (1): 27–44.

Brown, G. (1991), "The Evolution of the Term 'Mixed Mathematics'," *Journal of the History of Ideas*, 52 (1): 81–102.

Bracken, H. M. (1974), *Berkeley*, New York: St. Martin's Press.

Buckle, S. (2004), *Hume's Enlightenment Tract*, New York: Oxford University Press.

Busch, K. R. (2016), "Hume's Alleged Lapse on the Causal Maxim," *Hume Studies*, 42 (1–2): 89–112.

Butts, R. (ed.) (1986), *Kant's Philosophy of Physical Science*, Dordrecht: D. Reidel Publishing Company.

Carroll, J. W. (1990), "The Humean Tradition," *The Philosophical Review,* 99 (2): 185–219.

Carroll, J. W. (1994), *Laws of Nature*, Cambridge: Cambridge University Press.

Carroll, J. W. (2012), "Laws of Nature," in E. N. Zalta (ed.), *The Stanford Encyclopedia of Philosophy* (Spring 2012 Edition), http://plato.stanford.edu/archives/spr2012/entries/laws-of-nature/

Chambers, E. (1728), *Cyclopædia, or an Universal Dictionary of Arts and Sciences, 2 volumes, with the 1753 Supplement.* Digitized by the University of Wisconsin Digital Collections Center, http://uwdc.library.wisc.edu/collections/HistSciTech/Cyclopaedia

Clarke, D. M. (1982), *Descartes' Philosophy of Science*, Manchester: Manchester University Press.

Cohen, E. D. (1977), "Hume's Fork," *Southern Journal of Philosophy*, 15 (4): 443–55.

Cohen, I. B. (1980), *The Newtonian Revolution*, Cambridge: Cambridge University Press.

Cohen, I. B. (1999), "A Guide to Newton's *Principia*," in *Principia. The Mathematical Principles of Natural Philosophy*, 1–370, Berkeley: University of California Press.

Cohen, I. B. and G. E. Smith (2002), "Introduction," in I. B. Cohen and G. E. Smith (eds), *Cambridge Companion to Newton*, 1–32, Cambridge: Cambridge University Press.

Comte, A. (2012), *Cours de philosophie positive, leçons 46–51*, Paris: Hermann.

Cottrell, J. (2015), "David Hume: Imagination," in J. Fieser and B. Dowden (eds), *Internet Encyclopedia of Philosophy*, https://www.iep.utm.edu/hume-ima/

Cottrell, J. (2018), "Hume on Space and Time: A Limited Defense," in A. Coventry and A. Sager (eds), *The Humean Mind*, 83–95, New York: Routledge.

Coventry, A. (2006), *Hume's Theory of Causation: A Quasi-Realist Interpretation*, London and New York: Continuum.

Cuicciardini, N. (1998), "Did Newton Use His Calculus in the *Principia*?" *Centaurus* 40: 303–44.

Cushing, J. T. (1998), *Philosophical Concepts in Physics. The Historical Relation between Philosophy and Scientific Theories*, Cambridge: Cambridge University Press.

Daniel, S. H. (2007), "The Harmony of the Leibniz-Berkeley Juxtaposition," in P. Phemister and S. Brown (eds), *Leibniz and the English-Speaking World*, 163–80, Dordrecht: Springer.

De Caro, M. (1992), "Galileo's Mathematical Platonism," in G. Czermak (ed.), *Philosophy of Mathematics*, 1–9, Wien: Hoelder-Pichler-Tempsky.

De Haro, S. (2019), "Science and Philosophy: A Love–Hate Relationship," *Foundation of Science*, https://doi.org/10.1007/s10699-019-09619-2

Deng, N. (2018), "Time, metaphysics of," *Routledge Encyclopedia of Philosophy*, doi:10.4324/9780415249126-N123-3

De Pierris, G. (2001), "Hume's Pyrrhonian Scepticism and the Belief in Causal Laws," *Journal of the History of Philosophy*, 39 (3): 351–83.

De Pierris, G. (2006), "Hume and Locke on Scientific Methodology: The Newtonian Legacy," *Hume Studies*, 32 (2): 277–330.

De Pierris, G. (2015), *Ideas, Evidence and Method: Hume's Skepticism and Naturalism Concerning Knowledge and Causation*, Oxford: Oxford University Press.

de Regt, H. W. (2017), *Understanding Scientific Understanding*, New York: Oxford University Press.

Demeter, T. (2012), "Hume's Experimental Method," *British Journal for the History of Philosophy*, 20 (3): 577–99.

Demeter, T. (2017), *David Hume and the Culture of Scottish Newtonianism*, Boston: Brill.

Demeter, T., B. Láng, and D. Schmal (2015), "Scientiae," in M. Sgarbi (ed.), *Encyclopedia of Renaissance Philosophy*, 1–15, Springer, doi:10.1007/978-3-319-02848-4_266-1

Descartes, R. (1983), *Principles of Philosophy*, trans V. R. Miller and R. P. Miller, Dordrecht: Reidel.

Descartes, R. (2000), *Discours de la méthode*, D. Moreau (ed.), Paris: Librairie Générale Française.

DeWitt, R. (2010), *Worldviews: An Introduction to the History and Philosophy of Science*, Chichester: Wiley-Blackwell.

DiSalle, Robert (2006), *Understanding Space-Time. The Philosophical Development of Physics from Newton to Einstein*, New York: Cambridge University Press.

Dretske, F. I. (1977), "Laws of Nature," *Philosophy of Science*, 44 (2): 248–68.

Downing, L. (1995), "Berkeley's Case against Realism about Dynamics," in R. Muehlmann (ed.), *Berkeley's Metaphysics. Structural, Interpretive, and Critical Essays*, 197–214, University Park: Penn State University Press.

Downing, L. (2005), "Berkeley's Natural Philosophy and Philosophy of Science," in K. P. Winkler (ed.), *Cambridge Companion to Berkeley*, 230–64. New York: Cambridge University Press.

Dowden, B. (2018), "Time," *Internet Encyclopedia of Philosophy*, ISSN 2161-0002, http://www.iep.utm.edu/time/

Ducheyne, S. (2001), "Isaac Newton on Space and Time: Metaphysician or Not?", *Philosophica*, 67 (1): 77–114.

Ducheyne, S. (2009), "'Newtonian' Elements in Locke, Hume, and Reid, or: How Far Can One Stretch a Label?," in S. Snobelen (ed.), *Enlightenment and Dissent. Isaac Newton in the Eighteenth Century*, 25, 62–105.

Ducheyne, S. (2012), *The Main Business of Natural Philosophy. Isaac Newton's Natural-Philosophical Methodology*, Dordrecht: Springer.

Dunton, J. (1692), *The Young-Students-Library*, London.

Dunlop, K. (2012), "What Geometry Postulates: Newton and Barrow on the Relationship of Mathematics to Nature," in A. Janiak and E. Schliesser (eds), *Interpreting Newton: Critical Essays*, 69–101, New York: Cambridge University Press.

Durland, K. (1996), "Hume's First Principle, His Missing Shade, and His Distinctions of Reason," *Hume Studies*, 22 (1), 105–22.

Dyke, H. (2007), "Science and Philosophy: Making Time for Each Other," ABC podcast, transcript available on https://www.abc.net.au/radionational/programs/ockhamsrazor/science-and-philosophy-making-time-for-each-other/3290084#transcript. Last accessed July 15, 2019.

Earman, J. (1989), *World Enough and Space-Time. Absolute versus Relational Theories of Space and Time*, Cambridge, London: The MIT Press.

Einstein, A. (1923), "On the Electrodynamics of Moving Bodies," in H. Lorentz et al. (eds), *The Principle of Relativity. A Collection of Original Memoirs on the Special and General Theory of Relativity*, 35–65, W. Perret and G. B. Jeffery (trans), New York: Dover Publications.

Einstein, A. (1936), "Physics and Reality," *Journal of the Franklin Institute* 221 (3): 349–82.

Einstein, A. (1949a), "Autobiographical Notes," in P. A. Schilpp (trans, ed.), *Albert Einstein. Philosopher–Scientist*, 1–94, New York: MJF Books.

Einstein, A. (1949b), "Remarks Concerning the Essays Brought Together in this Co-Operative Volume," in P. A. Schilpp (trans, ed.), *Albert Einstein. Philosopher–Scientist*, 665–93, New York: MJF Books.

Einstein, A. (1981a), "Principles of Theoretical Physics," Inaugural address before the Prussian Academy of Sciences, 1914, in C. Seelig (ed.), *Ideas and Opinions*, 216–19, S. Bargmann (trans), New York: Dell Publishing.

Einstein, A. (1981b), "Principles of Research," Address delivered at a celebration of Max Planck's sixtieth birthday before the Physical Society in Berlin, 1918, in C. Seelig (ed.), *Ideas and Opinions*, 219–22, S. Bargmann (trans), New York: Dell Publishing.

Einstein, A. (1981c), "What Is The Theory of Relativity," written at the request of the London *Times*, November 28, 1919, in C. Seelig (ed.), *Ideas and Opinions*, 222–7, S. Bargmann (trans), New York: Dell Publishing.

Einstein, A. (1981d), "On the Theory of Relativity," Lecture at King's College, London, 1921, in C. Seelig (ed.), *Ideas and Opinions*, 240–3, S. Bargmann (trans), New York: Dell Publishing.

Einstein, A. (1981e), "On the Method of Theoretical Physics," The Herbert Spencer Lecture, delivered at Oxford, June 10, 1933, in C. Seelig (ed.), *Ideas and Opinions*, 263–70, S. Bargmann (trans), New York: Dell Publishing.

Einstein, A. (1981f), "The Problem of Space, Ether, and the Field in Physics," in C. Seelig (ed.), *Ideas and Opinions*, 270–8, S. Bargmann (trans), New York: Dell Publishing.

Einstein, A. (1981g), "Physics and Reality," in C. Seelig (ed.), *Ideas and Opinions*, 283–315, S. Bargmann (trans), New York: Dell Publishing.

Einstein, A. (1981h), "The Fundaments of Theoretical Physics," in C. Seelig (ed.), *Ideas and Opinions*, 315–26, S. Bargmann (trans), New York: Dell Publishing.

Einstein, A. (1981i), "Relativity and the Problem of Space," in C. Seelig (ed.), *Ideas and Opinions*, 350–66, S. Bargmann (trans), New York: Dell Publishing.

Einstein, A. (1981j), "Remarks on Bertrand Russell's Theory of Knowledge," in C. Seelig (ed.), *Ideas and Opinions*, 29–35, P. A. Schilpp (trans), New York: Dell Publishing.

Einstein, A. (1987), *The Collected Papers of Albert Einstein, Volume 1: The Early Years, 1879–1902*, (eds, trans) A. Beck and P. Havas, Princeton: Princeton University Press.

Einstein, A. (1998), *The Collected Papers of Albert Einstein, Volume 8: The Berlin Years: Correspondence, 1914–1918*, Ann M. Hentschel (trans), Princeton: Princeton University Press.

Einstein, A. (2001), *Relativity. The Special and General Theory*, R. W. Lawson (trans), London and New York: Routledge.

Einstein, A. (2002), *The Collected Papers of Albert Einstein, Volume 7: The Berlin Years: Writings, 1918–1921*, (ed.) M. Janssen, Princeton: Princeton University Press.

Einstein, A. (2003), *The Meaning of Relativity*, E. P. Adams, E. G. Strauss, and S. Bargmann (trans), London: Routledge.

Esfeld, M. (2011), "Causal Realism," in Dennis Dieks et al. (eds), *Probabilites, Laws, and Structures*, 157–168, Dordrecht: Springer.

Euclid (1945), *The Optics of Euclid*, H. E. Burton (trans), *Journal of the Optical Society of America*, 35 (5): 357–72.

Feyerabend, P. (1980), "Zahar on Mach, Einstein and Modern Science," *The British Journal for the Philosophy of Science*, 31 (3): 273–82.

Feyerabend, P. (1984), "Mach's Theory of Research and Its Relation to Einstein", *Studies in History and Philosophy of Science Part A*, 15 (1): 1–22.

Field, H. (2003), "Causation in a Physical World", in D. Zimmerman (ed.), *Oxford Handbook of Metaphysics*, 435–60, Oxford: Oxford University Press.

Fine, K. (2012), "What Is Metaphysics?", in T. Tahko (ed.), *Contemporary Aristotelian Metaphysics*, 8–25, New York: Cambridge University Press.

Flew, A. (1984), *A Dictionary of Philosophy*, New York: St. Martin's Press.

Force, J. E. (1987), "Hume's Interest in Newton and Science," *Hume Studies*, 23 (2): 166–216.

Freiberger, M. (2012), "Schrödinger's Equation—What Is It?", *Plus Magazine*, August 2, 2012, https://plus.maths.org/content/schrodinger-1

Gabbey, A. (1980), "Huygens and Mechanics," in H. J. M. Bos, M. J. S. Rudwick, H. A. M. Snelders, and R. P. W. Visser (eds), *Studies on Christiaan Huygens*, 166–99, Lisse: Swets & Zeitlinger.

Galileo, G. (1965), *Il Saggiatore*, S. Drake (trans), Milano: Feltrinelli.

Galison, P. (2003), *Einstein's Clocks, Poincaré's Maps*, New York and London: W. W. Norton & Company.

Galison, P. (2008), "Ten Problems in History and Philosophy of Science," *Isis*, 99, 111–24.

Garrett, D. (1997), *Cognition and Commitment in Hume's Philosophy*, New York: Oxford University Press.

Garrett, D. (2015a), "Hume's Theory of Causation: Inference, Judgment, and the Causal Sense," in D. C. Ainslie and A. Butler (eds), *The Cambridge Companion to Hume's Treatise*, 69–100, New York: Cambridge University Press.

Garrett, D. (2015b), *Hume*, New York: Routledge.

Garrett, D. (2018), "Hume's System of the Sciences," in A. Coventry and A. Sager (eds), *The Humean Mind*, 55–71, New York: Routledge.

Gaukroger, S. (2002), *Descartes' System of Natural Philosophy*, Cambridge: Cambridge University Press.

Gaukroger, S. (2010), *The Collapse of Mechanism and the Rise of Sensibility*, Oxford: Clarendon.

Gilmore, C., D. Costa and C. Calosi (2016), "Relativity and Three Four Dimensionalisms," *Philosophy Compass*, 11: 102–20.

Gossman, L. (1960), "Two Unpublished Essays on Mathematics in the Hume Papers," *Journal of the History of Ideas*, 21 (3): 442–9.

Grant, E. (2007), *A History of Natural Philosophy. From the Ancient World to the Nineteenth Century*, New York: Cambridge University Press.

Greene, B. (2004), *The Fabric of the Cosmos*, New York: Albert Knopf.

Grünbaum, A. (1973), *Philosophical Problems of Space and Time*, Dordrecht: D. Reidel.

Haakonssen, K. (2004), "The Idea of Early Modern Philosophy," in J. B. Schneewind (ed.), *Teaching New Histories of Philosophy*, Princeton: University Center for Human Values.

Hakkarainen, J. (2012a), "Hume as a Trope Nominalist," *Canadian Journal of Philosophy*, 42 (1): 55–66.

Hakkarainen, J. (2012b), "Hume's Scepticism and Realism," *British Journal for the History of Philosophy*, 20 (2): 283–309.

Hakkarainen, J. (2019), "Hume on the Unity of the Determinations of Extension," *Logical Analysis and History of Philosophy*, 22 (1): 219–33.

Harper, W. (2016), "Newton's Argument for Universal Gravitation," in R. Iliffe and G. E. Smith (eds), *Cambridge Companion to Newton*, 2nd edn, 229–60, Clays, St Ives: Cambridge University Press.

Harris, J. A. (2015), *Hume: An Intellectual Biography*, New York: Cambridge University Press.

Harrison, P. (2002), "Original Sin and the Problem of Knowledge in Early Modern Europe," *Journal of the History of Ideas*, 63 (2), 239–59.

Hatfield, G. (1996), "Was the Scientific Revolution Really a Revolution in Science?," in F. J. Ragep, S. P. Ragep, and S. Livesey (eds), *Proceedings of Two Conferences on Pre-modern Science Held at the University of Oklahoma*, 489–523, Leiden: Brill.

Hawking, S. and L. Mlodinow (2010), *The Grand Design*, New York: Bantam Books.

Hazony, Y. (2014), "Newtonian Explanatory Reduction and Hume's System of the Sciences," in Z. Biener and E. Schliesser (eds), *Newton and Empiricism*, 138–70, New York: Oxford University Press.

Hentschel, K. (1985), "On Feyerabend's Version of 'Mach's Theory of Research and Its Relation to Einstein," *Studies in History and Philosophy of Science Part A*, 16 (4): 387–94.

Hentschel, K. (1992), "Einstein's Attitude towards Experiments: Testing Relativity Theory 1907–1927," *Studies in History and Philosophy of Science*, 23 (4): 593–624.

Hight, M. A. (2010), "Berkeley's Metaphysical Instrumentalism," in Silvia Parigi (ed.), *George Berkeley: Religion and Science in the Age of the Enlightenment*, 15–29, Dordrecht: Springer.

Hobbes, Thomas (1985), 'Translation of Hobbes' Dialogus physicus' by Simon Schaffer in *Leviathan and the Air-Pump: Hobbes, Boyle and the Experimental Life*, 345–92, Princeton: Princeton University Press.

Holden, T. (2014), "Hume's Absolute Necessity," *Mind*, 123 (490): 377–413.

Holton, G. (1968), "Mach, Einstein, and the Search for Reality," *Daedalus*, 97 (2): 636–73.

Holton, G. (1992), *More on Mach and Einstein*, in J. T. Blackmore (ed.), *Ernst Mach—A Deeper Look: Documents and Perspectives*, 263–76, Dordrecht: Springer.

Howard, D. (1994), "Einstein, Kant, and the Origins of Logical Empiricism," in W. C. Salmon and G. Wolters (eds), *Language, Logic, and the Structure of Scientific Theories*, 45–105, Pittsburgh: University of Pittsburgh Press.

Howard, D. (2005), "Albert Einstein as a Philosopher of Science," *Physics Today*, December 2005, http://dx.doi.org/10.1063/1.2169442

Hurlbutt, R. H. (1965), *Hume, Newton, and the Design Argument*, Lincoln: The University of Nebraska Press.

Jammer, M. (1957), *Concepts of Force: A Study in the Foundations of Dynamics*, Cambridge, MA: Harvard University Press.

Janiak, A. (2007), "Newton and the Reality of Force," *Journal of the History of Philosophy*, 45 (1): 127–47.

Janiak, A. (2012), "Metaphysics and Natural Philosophy in Descartes and Newton," *Foundations of Science*, 18 (3): 403–17.

Janiak, A. (2013), "Isaac Newton," in P. R. Anstey (ed.), *Oxford Handbook of British Philosophy in the Seventeenth Century*, 96–115, Oxford: Oxford University Press.

Janiak, A. (2015), *Newton*, Malden, MA: Wiley-Blackwell.

Janssen, M. and C. Lehner (2014), "Introduction," in *The Cambridge Companion to Einstein*, 1–37, Cambridge: Cambridge University Press.

Jesseph, D. M. (1992), "Introduction" to Berkeley's *De Motu and the Analyst*, 3–44, Dordrecht: Kluwer Academic Publishers.

Jesseph, D. M. (1993), *Berkeley's Philosophy of Mathematics*, Chicago and London: University of Chicago Press.

Kail, P. (2007), *Projection and Realism in Hume's Philosophy*, Oxford: Oxford University Press.

Kant, I. (1998), *Critique of Pure Reason*, P. Guyer and A. W. Wood (trans), New York: Cambridge University Press.

Kervick, D. (2016), "Hume's Perceptual Relationism," *Hume Studies*, 42 (1–2): 61–87.

Kervick, D. (2018), "Hume's Colors and Newton's Colored Lights," *Journal of Scottish Philosophy*, 16 (1): 1–18.

Knight, R. D. (2008), *Physics: for Scientists and Engineers*, San Francisco: Pearson–Addison-Wesley.

Kochiras, H. (2011), "Gravity's Cause and Substance Counting: Contextualizing the Problems," *Studies in the History and Philosophy of Science*, 42 (1): 167–84.

Kuhn, T. (1996), *The Structure of Scientific Revolutions*, Chicago: University of Chicago Press.

Kuorikoski, J. (2014), "How to Be a Humean Interventionist," *Philosophy and Phenomenological Research*, 89 (2): 333–51.

Kuypers, M. S. ([1930]1966), *Studies in the Eighteenth Century Background of Hume's Empiricism*, New York: Russell & Russell.

Ladyman, J. and D. Ross (2007), *Every Thing Must Go: Metaphysics Naturalized*, New York: Oxford University Press.

Landy, D. (2018), *Hume's Science of Human Nature. Scientific Realism, Reason, and Substantial Explanation*, New York: Routledge.

Lehrer, K. (1978), "Why Not Scepticism?", in G. S. Pappas and M. Swain (eds), *Essays on Knowledge and Justification*, 346–63, Ithaca, New York: Cornell University Press.

Leibniz, G. W. (1989a), "Against Barbaric Physics: Toward a Philosophy of What There Actually Is and Against the Revival of the Qualities of the Scholastics and Cimerical Intelligences," in R. Ariew and D. Garber (eds, trans), *Philosophical Essays*, 312–20, Indianapolis & Cambridge: Hackett Publishing Company.

Leibniz, G. W. (1989b), "From the Letters to Clarke (1715–16)," in R. Ariew and D. Garber (eds, trans), *Philosophical Essays*, 321–46, Indianapolis & Cambridge: Hackett Publishing Company.

Leibniz, G. W. (1996), *New Essays Concerning Human Understanding*, (eds, trans), Peter Remnant and Jonathan Bennett, Glasgow: Cambridge University Press.

Lemos, N. (2010), "The Common Sense Tradition," in J. Dancy, E. Sosa, and M. Steup (eds), *A Companion to Epistemology*, 53–62, Malden, MA: Blackwell Publishing.

Lenzen, V. F. (1949), "Einstein's Theory of Knowledge," in P. A. Schilpp (ed.), *Albert Einstein. Philosopher Scientist*, 355–84, Library of living philosophers, La Salle, Illinois: Opern Court.

Lewis, D. K. (1973), *Counterfactuals*, Cambridge, MA: Harvard University Press.

Lowe, E. J. (2011), "The Rationality of Metaphysics," *Synthese*, 178 (1): 99–109.

MacDonald, C. (2005), *Varieties of Things: Foundations of Contemporary Metaphysics*, Malden, MA: Wiley-Blackwell.

Mach, E. (1919), *The Science of Mechanics*, T. McCormack (trans), Chicago, London: Open Court.

MacIntosh, J. J., and P. Anstey (2014), "Robert Boyle," in E. N. Zalta (ed.), *The Stanford Encyclopedia of Philosophy*, https://plato.stanford.edu/entries/boyle/

Macintyre, A. (1984), "The Relationship of Philosophy to Its Past," in R. Rorty, J. B. Schneewind, and Q. Skinner (eds), *Philosophy in History*, 31–48, Cambridge: Cambridge University Press.

Maclaurin, J. and H. Dyke (2012), "What Is Analytic Metaphysics For?", *Australasian Journal of Philosophy*, 90 (2): 291–306.

Malebranche, N. (1977), *The Search after Truth*, T. M. Lennon and P. J. Olscamp (eds, trans), Cambridge: Cambridge University Press.

Martin, M. (1993), "The Rational Warrant for Hume's General Rules," *Journal of the History of Philosophy*, 31 (2): 245–57.

Maudlin, T. (2007), *The Metaphysics Within Physics*, New York: Oxford University Press.

Maudlin, T. (2012), *Philosophy of Physics: Space and Time*. Princeton: Princeton University Press.

Maxwell, N. (2012), "In Praise of Natural Philosophy: A Revolution for Thought and Life," *Philosophia*, 40 (4): 705–15.

Maxwell, N. (2019), "Natural Philosophy Redux," *Aeon* online magazine, May 13, 2019. https://aeon.co/essays/bring-back-science-and-philosophy-as-natural-philosophy

Menzies, P. (2014), "Counterfactual Theories of Causation," in E. N. Zalta (ed.), *The Stanford Encyclopedia of Philosophy*, https://plato.stanford.edu/entries/causation-counterfactual/

McTaggart, J. M. E. (1908), "The Unreality of Time." *Mind*, 17, 457–73.

Mikkeli, H. (2018), "Giacomo Zabarella," in E. N. Zalta (ed.), *The Stanford Encyclopedia of Philosophy*, https://plato.stanford.edu/entries/zabarella/

Miller, D. M (2009), "Qualities, Properties, and Laws in Newton's Induction," *Philosophy of Science*, 76 (5): 1052–63.

Millican, P. (1998), "Hume on Reason and Induction: Epistemology or Cognitive Science?", *Hume Studies*, 24 (1): 141–60.

Millican, P. (2002), "Hume's Sceptical Doubts Concerning Induction," in P. Millican (ed.), *Reading Hume on Human Understanding: Essays on the first Enquiry*, 107–74, New York: Oxford University Press.

Millican, P. (2007), "Introduction" to David Hume's *An Enquiry Concerning Human Understanding*, ix–lxv, New York: Oxford University Press.

Millican, P. (2009), "Hume, Causal Realism, and Causal Science," *Mind*, 118 (471): 647–712.

Millican, P. (2017), "Hume's Fork, and his Theory of Relations," *Philosophy and Phenomenological Research*, 95 (1): 3–65.

Minkowski, H. (1923), "Space and Time," in H. Lorentz et al. (eds), *The Principle of Relativity. A Collection of Original Memoirs on the Special and General Theory of Relativity*, W. Perret and G. B. Jeffery (trans), 73–91, Dover Publications.

Morganti, M. and T. Tahko (2017), "Moderately Naturalistic Metaphysics," *Synthese*, 194 (7), 2557–80.

Morris, W. E. (2009), "Meaning(fullness) Without Metaphysics: Another Look at Hume's 'Meaning Empiricism'," *Philosophia*, 37 (3): 441–54.

Morris, W. E. (2011), "Hume's Epistemological Legacy," in E. S. Radcliffe (ed.), *A Companion to Hume*, Malden, Oxford, West Sussex: Wiley-Blackwell.

Morris, W. E. and C. R. Brown (2014), "David Hume," in E. N. Zalta (ed.), *The Stanford Encyclopedia of Philosophy*, https://plato.stanford.edu/entries/hume/

Nadler, S. (2000), "Malebranche on Causation," in S. Nadler (ed.), *The Cambridge Companion to Malebranche*, 112–38, Cambridge: Cambridge University Press.

Newman, L. (2014), "Descartes' Epistemology," in E. N. Zalta (ed.), *The Stanford Encyclopedia of Philosophy*, https://plato.stanford.edu/entries/descartes-epistemology/

Newton, I. (1974a), "From a Letter to Cotes," in H. S. Thayer (ed.), *Newton's Philosophy of Nature: Selections from his Writings*, 5–6, 7–8, New York: Hafner Press.

Newton, I. (1974b), "From a Letter to Oldenburg," in H. S. Thayer (ed.), *Newton's Philosophy of Nature: Selections from his Writings*, 6–7, New York: Hafner Press.

Newton, I. (1979), *Opticks*, Mineola: Dover Publications.

Newton, I. (1999), *Principia. The Mathematical Principles of Natural Philosophy*, I. B. Cohen, J. Budenz and A. Whitman (trans), Berkeley: University of California Press.

Newton, I. (2004), "De Gravitatione," in A. Janiak (ed.), *Newton: Philosophical Writings*, 12–39, Cambridge: Cambridge University Press.

Norton, D. F. (2006), "Editor's Introduction," in David Fate Norton and Mary J. Norton (eds), *A Treatise of Human Nature*, 19–199. New York: Oxford University Press.

Norton, J. D. (2003), "Causation as Folk Science," *Philosopher's Imprint*, 3 (4), 1–22.

Norton, J. D. (2010), "How Hume and Mach Helped Einstein to Find Special Relativity," in M. Domski et al. (eds), *Discourse on a New Method. Reinvigorating the Marriage of History and Philosophy of Science*, 359–87, Chicago and La Salle, IL: Open Court.

Norton, J. D. (2014), "Einstein's Special Theory of Relativity and the Problems in the Electrodynamics of Moving Bodies that Led him to It," in M. Janssen (ed.), *Cambridge Companion to Einstein*, 72–102, New York: Cambridge University Press.

Øhrstrøm, P. (1980), "Conventionality in Distant Simultaneity," *Foundations of Physics*, 10 (3–4): 333–43.

Oki, S. (2013), "The Establishement of 'Mixed Mathematics' and Its Decline 1600–1800," *Historia Scientiarum*, 23 (2): 82–91.

Ott, W. (2009), *Causation and Laws of Nature in Early Modern Philosophy*, New York: Oxford University Press.

Owen, D. (2002), "Hume versus Price on Miracles and Prior Probabilities: Testimony and the Bayesian Calculation," in P. Millican (ed.), *Reading Hume on Human Understanding. Essays on the First Enquiry*, 335–48, New York: Oxford University Press.

Owen, D. (1999), *Hume's Reason*, New York: Oxford University Press.

Palmerino, C. R. (2016), "Reading the Book of Nature: The Ontological and Epistemological Underpinnings of Galileo's Mathematical Realism," in G. Gorham, B. Hill, E. Slowik, and C. K. Waters (eds), *The Language of Nature: Reassessing the Mathematization of Natural Philosophy in the Seventeenth Century*, 29–50, Minneapolis: University of Minnesota Press.

Piercey, R. (2003), "Doing Philosophy Historically," *The Review of Metaphysics*, 56 (4), 779–800.

Pigliucci, M. (2013), "The Demarcation Problem: A (Belated) Response to Laudan," in M. Pigliucci and M. Boudry (eds), *Philosophy of Pseudoscience: Reconsidering the Demarcation Problem*, 9–28, Chicago: University of Chicago Press.

Pine, R. (1989), *Science and the Human Prospect*, Belmont, CA: Wadsworth.

Polkinghorne, J. (2002), *Quantum Theory: A Very Short Introduction*, New York: Oxford University Press.

Popper, K. (2002), *The Logic of Scientific Discovery*, London and New York: Routledge.

Psillos, S. (2002), *Causation and Explanation*, Stocksfield: Acumen.

Ramsey, F. P. (1978), "Universals of Law and Fact," in D. H. Mellor (ed.), *Foundations*, 128–32, London: Routledge & Kegan Paul.

Reichenbach, H. (1951), *The Rise of Scientific Philosophy*, Berkeley: University of California Press.

Ribeiro, C. (2015), "The Complementary of Science and Metaphysics," *Philosophica*, 90: 61–92.

Rini, A. (2016), "Aristotle on the Necessity of the Consequence," in M. Cresswell, E. Mares, and A. Rini (eds), *Logical Modalities from Aristotle to Carnap. The Story of Necessity*, 11–28, Cambridge: Cambridge University Press.

Robison, W. (2018), "Hume, Descartes, and Adam: Hume's Project," *Proceedings of the 45th Annual Hume Society Conference*, Budapest: Hungarian Academy of Sciences.

Rorty, R., J. B. Schneewind, and Q. Skinner (1984), "Introduction," in R. Rorty, J. B. Schneewind, and Q. Skinner (eds), *Philosophy in History*, 1–14, Cambridge: Cambridge University Press.

Rosenberg, A. (1993), "Hume's Philosophy of Science," in D. F. Norton (ed.), *The Cambridge Companion to Hume*, 64–89, New York: Cambridge University Press.

Rosier-Catach, I. (2016), *Adam, la nature humaine, avant et après. Épistémologie de la chute*, Publications de la Sorbonne: Paris.

Roux, S., D. Garber (2013), "Introduction," in S. Roux and D. Garber (eds), *The Mechanization of Natural Philosophy*, xi–xviii, Boston Studies in the Philosophy of Science, vol. 282, New York: Springer.

Russell, B. (1953), "On the Notion of Cause, with Application to the Free Will-Problem," in M. Brodbeck and H. Feigl (eds), *Readings in the Philosophy of Science*, 387–407, New York: Appleton-Century-Crofts.

Russell, P. (1997), "Clarke's 'Almighty Space' and Hume's Treatise," in J. Dybikowski (ed.), *Enlightenment and Dissent*, 83–113, Aberystwyth: The University of Wales.

Rutherford, D. (1992), "Leibniz's Principle of Intelligibility," *History of Philosophy Quarterly*, 9 (1): 35–49.

Rutherford, D. (2007), "Innovation and Orthodoxy in Early Modern Philosophy," in D. Rutherford (ed.), *Cambridge Companion to Early Modern Philosophy*, 11–38, New York: Cambridge University Press.

Ryan, T. (2003), "Hume's Argument for the Temporal Priority of Causes," *Hume Studies*, 29 (1): 29–41.

Rynasiewicz, R. (2014), "Newton's Views on Space, Time, and Motion," in E. N. Zalta (ed.), *The Stanford Encyclopedia of Philosophy*, https://plato.stanford.edu/entries/newton-stm/

Sapadin, E. (1997), "A Note on Newton, Boyle, and Hume's 'Experimental Method'," *Hume Studies*, 23 (2): 337–44.

Sapadin, E. (2009), "Newton, First Principles, and Reading Hume," *Archiv für Geschichte der Philosophie*, 74 (1), 74–104.

Schabas, M. (2001), "David Hume on Experimental Natural Philosophy, Money, and Fluids," *History of Political Economy*, 33 (3): 411–35.

Schabas, M. (2005), *The Natural Origins of Economics*, Chicago: University of Chicago Press.

Schafer, K. (2014), "Curious Virtues in Hume's Epistemology," *Philosophers Imprint*, 14: 1–20.

Schaffer, J. (2008), "Causation and Laws of Nature: Reductionism," in T. Sider, J. Hawthorne, and D. W. Zimmerman (eds), *Contemporary Debates in Metaphysics*, 82–107, Malden, MA: Blackwell.

Schliesser, E. (2007), "Hume's Newtonianism and anti-Newtonianism," in E. N. Zalta (ed.), *The Stanford Encyclopedia of Philosophy*, https://plato.stanford.edu/entries/hume-newton/

Schliesser, E. (2009), "Hume's Attack on Newton's Philosophy," in Stephen Snobelen (ed.), *Enlightenment and Dissent. Isaac Newton in the Eighteenth Century*, no. 25, 167–203.

Schliesser, E. (2011), "Without God: Gravity as a Relational Quality of Matter in Newton's Treatise," in D. Jalobeanu and P. R. Anstey (eds), *Vanishing Matter and the Laws of Motion. Descartes and Beyond*, 80–102, New York: Routledge.

Schliesser, Eric (2013), "Newton's Philosophy of Time," in A. Bardon and H. Dyke (eds), *A Companion to the Philosophy of Time*, 87–101, Chichester, West Sussex: Wiley-Blackwell.

Schmaltz, T. M. (2013), "What Has History of Science to Do with History of Philosophy," in M. Lærke, J. E. H. Smith, and E. Schliesser (eds), *Philosophy and Its History. Aims and Methods in the Study of Early Modern Philosophy*, 301–23, New York: Oxford University Press.

Sergeant, J. (1696), *Solid Philosophy ASSERTED, Against the FANCIES of the IDEISTS: OR, THE METHOD to SCIENCE Farther Illustrated. WITH Reflexions on Mr. LOCKE's ESSAY Concerning Human Understanding*, London: Roger Clavil.

Shapin, S. and S. Schaffer (1985), *Leviathan and the Air-Pump: Hobbes, Boyle and the Experimental Life*, Princeton: Princeton University Press.

Shapiro, A. (2004), "Newton's Experimental Philosophy," *Early Modern Science and Medicine*, 9 (3): 185–217.

Sklar, L. (1990), "Real Quantities and Their Sensible Measures," in P. Bricker and R. I. G. Hughes (eds), *Philosophical Perspectives on Newtonian Science*, 57–76, Cambridge, MA, London: MIT Press.

Slavov, M. (2013), "Newton's Law of Universal Gravitation and Hume's Conception of Causality," *Philosophia Naturalis: Journal for the Philosophy of Nature*, 52 (2): 277–305.

Slavov, M. (2016a), "Empiricism and Relationism Intertwined: Hume and Einstein's Special Theory of Relativity," *Theoria: An International Journal for Theory, History and Foundations of Science*, 31 (1): 247–63.

Slavov, M. (2016b), *Essays Concerning Hume's Natural Philosophy*, Doctoral Dissertation, Jyväskylä: Jyväskylä University Printing House.

Slavov, M. (2016c), "Hume on the Laws of Dynamics: The Tacit Assumption of Mechanism," *Hume Studies*, 42 (1–2): 113–36.

Slavov, M. (2016d), "Newtonian and non-Newtonian Elements in Hume," *Journal of Scottish Philosophy*, 14 (3): 275–96.

Slavov, M. (2017), "Hume's Fork and Mixed Mathematics," *Archiv für Geschichte der Philosophie*, 99 (1): 102–19.

Slavov, M. (2018), "Ajan havaitsemisesta: onko aika empiirinen käsite?" in M. Tuominen and H. Laiho (eds), *Havainto*, 233–40, Turku: Painosalama Oy.

Slavov, M. (2019), "Time Series and Non-reductive Physicalism," *KronoScope*, 19 (1), 25–38.

Slavov, M. (2020), "Universal Gravitation and the (Un)Intelligibility of Natural Philosophy," *Pacific Philosophical Quarterly* 101 (1), 129–57.

Slowik, E. (2005), "On the Cartesian Ontology of General Relativity: Or, Conventionalism in the History of the Substantival-Relational Debate," *Philosophy of Science*, 72 (5): 1312–23.

Slowik, E. (2017), "Descartes' Physics," in E. N. Zalta (ed.), *The Stanford Encyclopedia of Philosophy*, https://plato.stanford.edu/entries/descartes-physics

Smeenk, C. (forthcoming), "Cosmology and Physical Astronomy in Newton's General Scholium," draft submitted for a volume regarding Newton's General Scholium, S. Snobelen et al. (eds).

Smith, D. W. (2018), "Phenomenology," in E. N. Zalta (ed.), *The Stanford Encyclopedia of Philosophy*, https://plato.stanford.edu/entries/phenomenology/

Smith, G. E. (2004), "The Methodology of the Principia," in I. B. Cohen and G. E. Smith (eds), *The Cambridge Companion to Newton*, 138–73, New York: Cambridge University Press.

Smith, G. E. (2007), "Isaac Newton," in E. N. Zalta (ed.), *The Stanford Encyclopedia of Philosophy*, https://plato.stanford.edu/entries/newton/

Smolin, L. and R. M. Unger (2015), *The Singular Universe and the Reality of Time: A Proposal in Natural Philosophy*, Cambridge: Cambridge University Press.

Snelders, H. A. M. (1989), "Christiaan Huygens and Newton's Theory of Gravitation," *Notes and Records of the Royal Society of London*, 43 (2): 209–22.

Snow, C. P. ([1959]1998), *The Two Cultures*, New York: Cambridge University Press.

Sorensen, Roy (2017), "Nothingness," in E. N. Zalta (ed.), *The Stanford Encyclopedia of Philosophy*, https://plato.stanford.edu/entries/nothingness/

Speziali, P. (1972), *Albert Einstein, Michelle Besso. Correspondence 1903–1955*, Paris: Herman.

Stachel, J. (1989), "'What Song the Syrens Sang': How Did Einstein Discover Special Relativity?" Italian translation in U. Curi (ed.), *L'Opera di Einstein*, 21–37, Ferrara: Gabriele Gorbino. Reprinted in Stachel, J. (2002), *Einstein from "B" to "Z,"* Boston: Birkhäuser.

Stachel, J. (2002), *Einstein from "B" to "Z"*, Boston: Birkhäuser.

Stanistreet, P. (2002), *Hume's Scepticism and the Science of Human Nature*, Aldershot: Ashgate.

Stein, H. (1970), "On the Notion of Field in Newton, Maxwell, and Beyond," in R. H. Stuewer (ed.), *Historical Perspectives of Science*, 264–87, Minneapolis: University of Minnesota Press.

Steiner, M. (1987), "Kant's Misrepresentation of Hume's Philosophy of Mathematics in the Prolegomena," *Hume Studies*, 13 (2): 400–10.

Stove, D. (1973), *Probability and Hume's Inductive Skepticism*, Oxford: Clarendon Press.

Swartz, N. (1995), "A Neo-Humean Perspective: Laws as Regularities," in F. Weinert (ed.), *Laws of Nature: Essays on the Philosophical, Scientific and Historical Dimensions*, 67–91, Berlin: de Gruyter.

Tahko, T. (2015), "The Modal Status of Laws: In Defence of a Hybrid View," *Philosophical Quarterly*, 65 (260): 509–28.

Traiger, S. (2018), "Hume on the Methods and Limits of the Science of Human Nature," in P. A. Reed and R. Vitz (eds), *Hume's Moral Philosophy and Contemporary Psychology*, 243–62, New York and London: Routledge.

Tooley, M. (1977), "The Nature of Law," *Canadian Journal of Philosophy*, 7 (4): 667–98.

Tooley, M. (1987), *Causation: A Realist Approach*, Oxford: Clarendon Press.

Tooley, M. (2004), "Probability and Causation," in P. Dowe and P. Noordhof (eds), *Cause and Chance: Causation in an Indeterministic World*, 77–119, New York: Routledge.

Twardy, C. (2014), *Causation, Causal Perception, and Conservation Laws*, Master's Thesis in History and Philosophy of Science, Indiana University.

van Fraassen, B. (1989), *Laws and Symmetry*, Oxford: Clarendon Press.

van Roomen, A. (1602), *Universiae mathesis idea*, Würzburg: Apud Georgium Fleischmann.

Vanzo, A. (2016), "Empiricism and Rationalism in Nineteenth-Century Histories of Philosophy," *Journal of the History of Ideas*, 77 (2): 253–82.

Visser, M. R. (2011), *An Epistemological Approach to Einstein's Thought Experiments*, Bachelor thesis in physics and astronomy, Institute for Theoretical Physics: University of Amsterdam.

Waismann, F. (2011), *Causality and Logical Positivism*, McGuinnes, B. F. (ed.), Dordrecht: Springer.

Weinert, F. (2005), "Einstein and Kant," *Philosophy*, 80 (4): 585–93.

Westfall, R. (1993), *The Life of Isaac Newton*, New York: Cambridge University Press.

Wilson, C. (2008), "From Limits to Laws: Origins of the 17th Century Conception of Nature as Legalite," in L. Daston and M. Stolleis (eds), *Natural Law and Laws of Nature in Early Modern Europe: Jurisprudence, Theology, Moral and Natural Philosophy*, 13–28, London and New York: Routledge.

Wittgenstein, L. (1922), *Tractatus Logico-Philosophicus*, London: Routledge & Kegan Paul.

Wittgenstein, L. (1958), *Philosophical Investigations*, Hoboken, New Jersey: Blackwell.

Woleński, J. (2004), "The History of Epistemology," in I. Niiniluoto, M. Sintonen, and J. Woleński (eds), *Handbook of Epistemology*, 3–56, Dordrecht, Boston, London: Kluwer Academic Publishers.

Wolters, G. (2012), "Mach and Einstein, or, Clearing Troubled Waters in the History of Science," in Christoph Lehner et al. (eds), *Einstein and the Changing World View of Physics*, 39–60, New York: Springer.

Wright, J. P. (1983), *The Sceptical Realism of David Hume*, Manchester: Manchester University Press.

Zahar, E. (1977), "Mach, Einstein, and the Rise of Modern Science," *The British Journal for the Philosophy of Science*, 28 (3): 195–213.

Index

This index includes historical figures (excluding Hume, Newton, and Einstein, as references to them are abundant), Hume scholars and key concepts.

Ingram Content Group UK Ltd.
Milton Keynes UK
UKHW020206210423
420545UK00004B/165